AF349271

5G/5G-Advanced, Wi-Fi 6/7, and Bluetooth 5/6

Douglas H Morais

5G/5G-Advanced, Wi-Fi 6/7, and Bluetooth 5/6

A Primer on Smartphone Wireless Technologies

Second Edition

 Springer

Douglas H Morais
San Mateo, CA, USA

ISBN 978-3-031-82829-4   ISBN 978-3-031-82830-0   (eBook)
https://doi.org/10.1007/978-3-031-82830-0

Original Springer Publication

This Springer imprint is published by the registered company Springer Nature Switzerland AG
The registered company address is: Gewerbestrasse 11, 6330 Cham, Switzerland

If disposing of this product, please recycle the paper.

Preface

This second edition is an update to the first edition of this text. Material in the first edition addressed 5G primarily as defined in 3GPP Releases 15 and 16, Wi-Fi as per Version 6, and Bluetooth as per Version 5. This edition adds to the body of information previously presented by including relevant material from 3GPP Releases 17 and 18 (5G-Advanced), Wi-Fi 7, and Bluetooth 6.

Wireless communications in today's world is diverse and ubiquitous. However, its largest and most visible segment is arguably cellular mobile networks. Communication between a mobile unit and a cellular network base station is commonly referred to as *mobile access*, and today most mobile units are so-called smartphones. A smartphone is much more than a phone, however. It's a portable computer device that combines mobile telephone communications with a host of computing functions. Further, in addition to mobile access, it incorporates Wi-Fi, Bluetooth, and GPS navigation wireless communication technologies. By including Wi-Fi, it avails itself, particularly in indoor environments, of high-speed Internet access independent of what can at times be costly and slow mobile access. This can be very helpful in high-density locations such as airports, shopping malls, sporting arenas, etc. Further, it allows the smartphone to act as a "hotspot," allowing one to connect devices such as laptops and tablets to the Internet via the smartphone. By including Bluetooth, it allows very short-range wireless communication between the smartphone and a range of devices such as speakers, headphones, hearing aids, and smartwatches. By including GPS navigation, it allows precise device location. It is the aim of this text to convey at a high level, and in a tractable fashion, the key technologies behind the latest versions of mobile assess, Wi-Fi, and Bluetooth wireless communication, these versions being 5G/5G-Advanced, Wi-Fi 6/7, and Bluetooth 5/6, respectively.

Texts already exist that individually focus on one of the three communication systems covered herein. No single one, however, as best the author can ascertain, addresses at a high level all three systems. Most, in general, cover these systems at a level that presupposes that the reader is already somewhat familiar with them and seeks a deeper understanding. This book, on the other hand, presupposes only a limited technical background in telecommunications and possibly no knowledge of some or all of the specific technologies covered.

The material presented herein commences, after a brief introduction, with an explanation of the key technologies employed by the communication systems under study, then addresses individually how each such system employs these technologies to create a working system. It is directed to industry professionals, academics, and members of the general public with a desire to learn more about smartphone wireless communications. It is intentionally not mathematically rigorous so as to be friendly to a wide audience.

The reader who is fairly familiar with the individual technologies covered in the earlier chapters and incorporated into the targeted systems covered in later chapters may choose to go directly to those later chapters, namely Chaps. 8, 9, and 10. In each of these chapters, where a specific covered technology is mentioned, reference is made to the section in a preceding chapter where the specific technology is described. Thus, should the need arise to refresh one's knowledge on a given technology, the location of its description is readily available.

For those desiring to explore some or all of the material presented in greater detail, several references are provided. Further, a number of texts currently exist, and more will no doubt continue to be released, that collectively treat the technologies presented here in greater detail. A goal of this text is to allow the reader, if he or she so desires, to tackle these more advanced books with confidence.

San Mateo, CA, USA

Douglas H. Morais

Contents

About the Author

Douglas H. Morais has decades of experience in the wireless communications field that encompasses product design, engineering management, executive management, consulting, and short course lecturing. He holds a B.Sc. from the University of Edinburgh, Scotland, an M.Sc. from the University of California, Berkeley, and a Ph.D. from the University of Ottawa, Canada, all in Electrical Engineering. Additionally, he is a graduate of the AEA/Stanford Executive Institute, Stanford University, California, is a Life Senior member of the IEEE, and a member of the IEEE Communications Society.

Dr. Morais has authored several papers on wireless digital communications, holds three US patents, one on point-to-multipoint wireless communications and two on digital modulation, and in addition to this book has authored six others: *Fixed Broadband Wireless Communications*, Prentice Hall PTR, 2004; *Key 5G Physical Layer Technologies,* Springer, 2020; *5G and Beyond Wireless Transport Technologies*, Springer, 2021; *Key 5G Physical Layer Technologies, Second Edition,* Springer, 2022; *5G NR, Wi-Fi 6, and Bluetooth LE 5*, Springer, 2023; and *Key 5G/5G-Advanced Physical layer Technologies, Third Edition,* Springer, 2024.

Abbreviations

2G	Second generation
3G	Third generation
3GPP	Third generation partnership project
4G	Fourth generation
5G	Fifth generation
AP	Access point
APP	Application
ARQ	Automatic repeat request
ATT	Attribute
BCC	Binary convolution coding
BER	Bit error rate
BIG	Broadcast isochronous group
BLE	Bluetooth low energy
BPSK	Binary phase shift keying
BR	Basic rate
BS	Base station
BSS	Basic service set
CA	Carrier aggregation
CBF	Coordinated beamforming
CIG	Connected isochronous group
CIS	Connected isochronous stream
CN	Check node
C-OFDMA	Coordinated OFDMA
CP	Control plane
CP	Cyclic prefix
CP-OFDM	Cyclic prefix OFDM
CRC	Cyclic redundancy check
CSR	Coordinated spatial reuse
DCM	Dual carrier modulation
DFT	Discrete Fourier transform
DFTS	Discrete Fourier transform spread

DL	Downlink
DSBSC	Double-sideband suppressed carrier
EDR	Enhanced data rate
EIRP	Equivalent isotropically radiated power
eMBB	Enhanced mobile broadband
eNB	eNodeB
EPC	Evolved packet core
ETSI	European Telecommunications Standard Institute
FDD	Frequency division duplexing
FEC	Forward error correction
FFT	Fast Fourier transform
FHSS	Frequency hopping spread spectrum
GAP	Generic access profile
GATT	Generic attribute
GFSK	Gaussian frequency shift keying
GI	Guard interval
gNB	gNodeB
GSM	Global system for mobile communication
HARQ	Hybrid automatic repeat request
IAB	Integrated access and backhaul
IDFT	Inverse discrete Fourier transform
IFFT	Inverse fast Fourier transform
IP	Internet protocol
ISI	Intersymbol interference
ITU	International Telecommunication Union
JTX	Joint transmission
L2CAP	Logical link control and adaptation protocol
LDPC	Low-density parity check
LE	Low energy
LL	Link layer
LLC	Logical link control
LLR	Log likelihood ratio
LOS	Line-of-sight
LTE	Long term evolution
MAC	Medium access control
MIC	Message integrity check
MIMO	Multiple-input multiple-output
MISO	Multiple-input single-output
MLO	Multi-link operation
mMIMO	Massive MIMO
mMTC	Massive machine-type communications
MT	Mobile termination
MTU	Maximum transfer unit
MU	Mobile unit
MU	Multi-user

Multiple-TRP	Multiple transmit/receive point
NCR	Network controlled repeater
NLOS	Non line-of-sight
NR	New radio
NTN	Non-terrestrial network
OFDM	Orthogonal frequency division multiplexing
OFDMA	Orthogonal frequency division multiple access
PAM	Pulse amplitude modulation
PAPR	Peak-to-average power ratio
PBR	Phase-based timing
PCM	Parity check matrix
PDCP	Packet data convergence protocol
PDSCH	Physical downlink shared channel
PDU	Protocol data unit
PHY	Physical
PLCP	Physical layer convergence procedure
PMD	Physical medium dependent
PMP	Point-to-multipoint
PRB	Physical resource block
PUSCH	Physical uplink shared channel
QAM	Quadrature amplitude modulation
QC	Quasi-cyclic
QoS	Quality of service
QPSK	Quadrature phase shift keying
RAN	Radio access network
RB	Resource block
RLC	Radio link control
RRC	Radio resource control
RTT	Round-trip timing
RU	Remote unit
RU	Resource unit
SD	Spatial diversity
SDAP	Service data adaptation protocol
SDU	Service data unit
SIG	Bluetooth Special Interest Group
SIMO	Single-input multiple-output
SINR	Signal to interference and noise ratio
SISO	Single-input single-output
SL	Sidelink
SM	Spatial multiplexing
SMP	Security management protocol
SNR	Signal-to-noise ratio
SR	Spatial reuse
SS	Spatial stream
STA	Station

STBC	Space-time block coding
SU	Single user
TCP	Transmission control protocol
TDD	Time division duplexing
TTI	Transmission time interval
TWT	Target wake time
UDP	User datagram protocol
UE	User equipment
UL	Uplink
UP	User plane
URLLC	Ultra-reliable and low-latency communication
V2X	Vehicle-to-anything
VN	Variable node

Chapter 1
5G/5G-Advanced, Wi-Fi 6/7, and Bluetooth 5/6 Introduction

1.1 Smartphone Wireless Communications

Wireless communications in today's world is diverse and ubiquitous. However, its largest and most visible segment is arguably cellular mobile networks. Originally, these networks provided voice communication only over analog networks. However, they have, over time, evolved into all digital networks providing more and more data capacity, to the point that the latest such networks communicate the so-called packet-switched data only. Now voice, video, and other applications are all converted to data and integrated into a common packet switched data stream. In such mobile networks, the coverage area consists of multiple adjacent cells, with each cell containing a fixed *base station*. See Fig. 1.1. Transmission to and from an individual *mobile unit* is normally between that unit and the base station that provides the best communication (normally but not always the closest one), and the coverage area around the base station can extend to several kilometers. Such communication is commonly referred to as *mobile access* and today most mobile units are the so-called *smartphones*, the most iconic of which is arguably the Apple iPhone.

A smartphone is much more than a phone. It is a portable computing device that combines mobile telephone communications with a host of computing functions. Furthermore, in addition to mobile access, it incorporates Wi-Fi, Bluetooth, GPS navigation, and in the case of the latest Apple iPhones, messaging via satellite communication technologies. By including Wi-Fi, it avails itself, particularly in indoor environments, of high-speed Internet access independent of what can at times be costly and slow mobile access. This can be very helpful in high-density locations such as airports, shopping malls, and sporting arenas. Furthermore, it allows the smartphone to act as a "hotspot," allowing one to connect devices such as laptops and tablets to the Internet via the smartphone. By including Bluetooth, it allows very short-range wireless communication between the smartphone and a range of devices such as speakers, headphones, hearing aids, and smart watches. By

© The Author(s), under exclusive license to Springer Nature
Switzerland AG 2025
D. H. Morais, *5G/5G-Advanced, Wi-Fi 6/7, and Bluetooth 5/6*,
https://doi.org/10.1007/978-3-031-82830-0_1

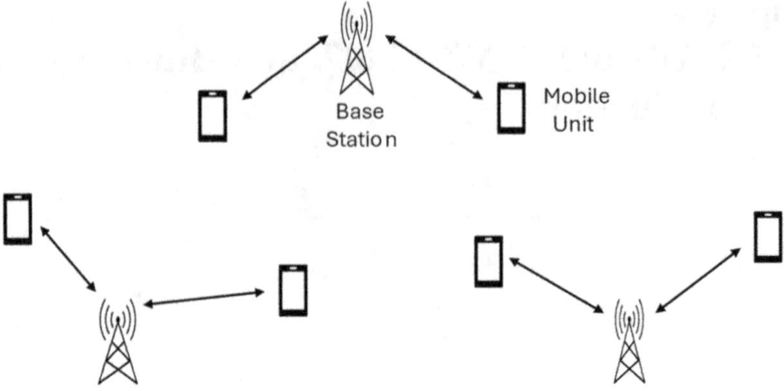

Fig. 1.1 Mobile network

including GPS navigation, it allows precise device location. By including satellite messaging technology, an iPhone allows its user to text messages while off the grid with no cellular or Wi-Fi coverage. It is the aim of this text to convey at a high level and in an easily understood fashion, the latest technologies behind mobile assess, Wi-Fi, and Bluetooth wireless communication. First, however, presented below are brief histories mobile access, Wi-Fi, and Bluetooth as well as a cursory look at GPS and iPhone satellite messaging technology.

1.2   Brief History of Mobile Access

The first commercial cellular telephone systems were introduced in the early 1980s, used analog technology, and are now referred to as *First-Generation* (1G) systems. They were designed primarily for the delivery of voice services.

Second-Generation (2G) systems appeared in the early 1990s. Though, like 1G systems, they were primarily aimed at voice services, they utilized digital modulation, allowing a higher voice capacity and support of low-rate data applications. Examples of 2G systems include the European designed *Global System for Mobile Communications* (GSM) system, the American designed IS-136 TDMA and IS-95 CDMA systems, and the Japanese designed PDC system. As an example of data capabilities, the original GSM standard supported circuit-switched data at 9.6 kb/s. By the late 1990s, however, with the introduction of *Enhanced Data Rate for GSM Evolution* (EDGE), user rates of between 80 and 120 kb/s were supported.

The process of broadly defining 3G standards for worldwide application was started by the International Telecommunications Union (ITU), which referred to such systems as *International Mobile Telecommunications 2000* (IMT-2000) systems. 3G systems became available in the early 2000s and represented a significant leap over 2G ones. These systems were conceived to deliver a wide range of services, including telephony, higher speed data than available with 2G, video, paging, and messaging.

The European Telecommunications Standard Institute (ETSI) was initially responsible for standardization of an IMT-2000 compliant system to be called *Universal Mobile Telecommunications System* (UMTS) and to be an evolution of GSM. However, in 1998, the *Third-Generation Partnership Project* (3GPP) was created with the mandate to continue this standardization work under the auspices of not only ETSI but also under those of other regional standardization development organizations, thus making the project a more global effort.

At about the same time that 3GPP was being created, a similar organization called 3GPP2 was being created under the auspices of North American, Japanese, and Chinese telecommunication associations. Like 3GPP, its goal was to standardize on an IMT-2000 compliant system.

All the 3G systems standardized by 3GPP and 3GPP2 utilized *Code Division Multiple Access* (CDMA) technology, provided significant increase in voice capacity, and offered much higher data rates over both circuit-switched and packet-switched bearers. Examples of early 3G systems include (a) the first release of UMTS, Release 99, referred to as WCDMA, (b) the first 3GPP2 release called CDMA 2000, and (c) a later release of 3GPP2 called EV-DO. These systems initially all provided peak *downlink* (DL) and *uplink* (UL) rates in the high kb/s to the very low Mb/s rates. These rates increased over time, but the breakout technology was UMTS whose continued evolution has led to extremely high data rates. Its CDMA-based Release 7, labeled HSPA+, with first service in 2011, provided a maximum DL rate of 84 Mb/s and a maximum UL rate of 23 Mb/s. Under its Release-8, however, which took place in December, 2008, it released not only a CDMA-based update but also an alternative system based on *orthogonal frequency division multiplexing* (OFDM) technology. This alternative system was labeled *Long-Term Evolution* (LTE). It was designed to permit a DL maximum data rate of 300 Mb/s and an UL maximum data rate of 75 Mb/s and commenced service in 2009.

The ITU commenced work on 4G systems in 2005 and labeled such systems *IMT-Advanced*. Such systems were defined as those capable of providing a maximum DL data rate of 1 Gb/s under low mobility conditions and of 100 Mb/s under high mobility conditions. 3GPP's candidate for IMT-Advanced was *LTE-Advanced* which was embodied in its Rel-10, which took place in March 2011. The IEEE's candidate was 802.16 m called *WirelessMAN-Advanced*. Both these candidates were based on OFDM technology. In October 2010, the ITU announced that LTE-Advanced and WirelessMAN-Advanced had been accorded the official designation of IMT-Advanced, qualifying them as true 4G technologies. However, as the UMTS evolved products had established such an overwhelming market position by that time, WirelessMAN-Advanced failed to take off, leaving LTE-Advanced as the only true 4G technology commercially available, with deployments commencing in the early 2010s.

The ITU in 2015 defined Fifth-Generation (5G) systems as those that meet its IMT-2020 requirements. IMT-2020 envisages the support of many usage scenarios, three of which it identified: enhanced mobile broadband (eMBB), ultra-reliable and low-latency communications (URLLC), and massive machine-type communications (mMTC):

- eMBB is the natural evolution of broadband services provide by 4G networks. It addresses human-centric use cases and applications, enabling the higher data rates and data volumes required to support ever-increasing multimedia services.
- URLLC addresses services requiring very low latency and very high reliability. Examples are factory automation, self-driving automobiles, and remote medical surgery.
- mMTC is purely machine centric, addressing services that provide connectivity to a massive number of devices, driven by the growth of the Internet of Things (IoT). It is expected the mMTC will support connection densities of up to one million devices per square kilometer. mMTC devices are expected to communicate only sporadically and then only a small amount of data. Thus, support of high data rates here is of less importance but low cost very important.

It should be noted that there are no hard boundaries between these stated scenarios. There will be many use cases that use less of the stated features or do not strictly fit into one of these cases but rather draw on the features of more than one scenario.

Among the many IMT-2020 envisaged requirements are (a) the capability of providing a peak DL data rate of 20 Gb/s and a peak UL data rate of 10 Gb/s, (b) user experienced DL data rates of up to 100 Mb/s and UL data rates of up to 50 Mb/s, (c) over the air latency of 1 ms, and (d) operation during mobility of up to 500 km/s.

3GPP commenced standardization works on its 5G wireless access technology in 2016 and labeled it *New Radio* (NR). The first release, Release 15, took place in June 2019. The first version of Release 15, ver. 15.1.0, allowed operation in two broad frequency ranges: sub-6 GHz (450 MHz to 6 GHz), referred to as FR1, and millimeter wave (from 24.25 to 52.6 GHz), referred to as FR2. Release 15, ver. 15.5.0, expanded FR1's range (410 MHz to 7.125 GHz) while leaving the FR2 range unchanged. Networks with performance approaching that of 5G NR Release 15 specifications began operating in 2019. The next release, Release 16, took place in July 2020, adding new capabilities and enhancements. The following release, Release 17, took place in June 2022. It provided significant advancements to features offered and allowed operation in the newly defined FR2–2 band which covers from 52.6 to 71 GHz (The original FR2 band was renamed FR2–1).

The next 3GPP release, Release 18, was released in June 2024, marks the start of what 3GPP has chosen to call *5G-Advanced* as together with Releases 16 and 17 it represents significant advancement in the technology and usage cases supported compared to Release 15.

1.3   A Brief History of Wi-Fi

Wi-Fi is officially a collection of wireless network protocols, based on the IEEE 802.11 series of standards and used commonly for Internet access, which is how it is used with smartphones, as well as for local area networking.

In 1991, AT&T and NCR invented the forerunner to 802.11, a system named WaveLAN and intended for use with cashier systems.

Six years later, in 1997, the first version of 802.11 was released. It operated in the 2.4-GHz unlicensed band and supported data rates of 1 and 2 Mb/s but is now obsolete.

802.11b (not 802.11a!) was released in 1999. It increased the maximum data rate to 11 Mb/s. Also, in 1999, the Wi-Fi Alliance was formed as a non-profit trade association to hold the Wi-Fi trademark under which most 802.11 type products are sold.

802.11a was originally outlined as clause 17 of the 1999 specification but was later defined in clause 18 of the 2012 specification. It supports data rates of from 1.5 to 54 Mb/s and operates in the 5 GHz unlicensed band.

Following the 1999 release, there have been numerous releases, all increasing the maximum data rate achievable or increasing the frequency bands supported. The most important of these releases are summarized in Table 1.1. One will notice the Wi-Fi Generation column. Historically, Wi-Fi equipment has simply listed the name of the IEEE standard that it supports. However, in 2018, the Wi-Fi Alliance introduced generational numbering to indicate equipment that it supports going back as far as 802.11n.

Wi-Fi 6/6E is designed to facilitate high-density wireless access and high-capacity wireless services, such as in high-density sports arenas, indoor high-density offices, large -scale outdoor public venues, etc. Compared to Wi-Fi 5, Wi-Fi 6/6E employs technologies that allow it to achieve a fourfold increase in average user throughput, to increase the number of concurrent users more than threefold and to increase in the maximum theoretical rate by 38%.

Table 1.1 Wi-Fi standard releases

Generation	IEEE standard	Release date	Frequency band/s (GHz)	Bandwidth (MHz)	Max. Data Rate (Mb/s)
–	802.11-1997	1997	2.4	22	2
–	802.11b	1999	2.4	22	11
–	802.11a	1999	5	20	54
–	802.11g	2003	2.4	20	54
Wi-Fi 4	802.11n High Throughput (HT)	2009	2.4/5	20/40	600
Wi-Fi 5	802.11ac Very High Throughput (VHT)	2013	5	20/40/80/160	6930
Wi-Fi 6	802.11ax High Efficiency Wireless (HEW)	2019	2.4/5	20/40/80/160	9600
Wi-Fi 6E	802.11ax (HEW)	2020	2.4/5/6	20/40/80/160	9600
Wi-Fi 7	802.11be Extremely High Throughput (EHT)	2024	2.4/5/6	20/40/80/160/240/320	46,100

Wi-Fi 7 builds on Wi-Fi 6, increasing available maximum bandwidth and modulation complexity, resulting in a theoretical maximum data rate of 46 Gb/s.

Going forward our interest will be in Wi-Fi 6/6E and Wi-Fi 7, these being the latest standards available. As alluded to above, the capabilities of Wi-Fi 6/6E and Wi-Fi 7 devices over Wi-Fi 5 ones are substantial and Wi-Fi 6/6E and Wi-Fi 7 devices are fully backward compatible with Wi-Fi 5 and older Wi-Fi devices.

1.4   A Brief History of Bluetooth

Bluetooth is a short-range wireless technology used for the exchange of data between devices, fixed or mobile, over short distances (a few meters up to about 240 m). It operates in the 2.4-GHz unlicensed band and is primarily used as an alternative to wired connections to link, for example, smartphones to a range of devices such as to headphones, hearing aids, speakers, and smart watches.

The Bluetooth standard was originally conceived at Ericsson back in 1994 and was named for a famous Viking king who united Denmark and Norway in the tenth century. Though it shares the 2.4-GHz band with Wi-Fi, it has always been intended as a much shorter range, lower data capacity, and lower power alternative to Wi-Fi.

In 1998, the *Bluetooth Special Interest Group* (SIG) was formed to publish and promote the standard and its continued revisions. Initially, it included only Ericsson, Nokia, Intel, and Toshiba, but currently contains over 36,000 members.

Following is a timeline of some of the more important Bluetooth events:

1999: **Bluetooth 1.0**, the first Bluetooth specification, released. It supports a theoretical maximum data rate of up to 1.0 Mb/s but in reality, more like 0.7 Mb/s. This rate structure is referred to as *basic rate* (BR). Coverage provided is about 10 m in an unobstructed environment.

2004: **Bluetooth 2.0** released. In addition to BR, it provides what's termed *Enhanced Data Rate* (EDR), theoretically rates of 2 and 3 Mb/s, but in reality, more like 1.4 and 2.1 Mb/s, respectively. Also, it increases the range to 30 m.

2009: **Bluetooth 3.0** released. As a purely Bluetooth link its realizable maximum data rate remains unchanged from that of version 2.0, i.e., 3 Mb/s, and coverage remains at 30 m. However, it offers a non-mandatory option called Bluetooth 3.0 + HS (HS for high speed) that provides a theoretical maximum data rate of 24 Mb/s, achieving this by allowing devices to transfer files over Wi-Fi while still communicating over Bluetooth. This much higher speed allows it to transmit larger amounts of audio and video data. Note that Bluetooth HS is only supported by Bluetooth interfaces marked as +HS. The main drawback of Bluetooth 3.0 is high power consumption that quickly drains the batteries of Bluetooth enabled devices.

2010: **Bluetooth 4.0** released. Bluetooth versions 1.0–3.0 are collectively referred to as *Bluetooth Classic*. However, Bluetooth 4.0 and all subsequent versions contain a "Core Configuration" referred to as *Bluetooth Low Energy* (BLE). With

BLE, the power consumption is considerably reduced. As a result, coin-cell battery operated devices such as fitness wearables and hearing aids become feasible. The BLE theoretical data rate is 1 Mb/s, but range increased to 60 m. Bluetooth 4 specifications also include the Bluetooth Classic configurations, i.e., BR, EDR, and HS.

2016: **Bluetooth 5.0** released. Here, the BLE theoretical rates can be 0.125, 0.5, 1, or 2 Mb/s with corresponding range of approximately 240, 120, 60, and 50 m, respectively. As with 4.0, it also offers BR, EDR, and HS.

2019: Bluetooth 5.1 released. Added direction finding capability an a number of enhancements.

2019: Bluetooth 5.2 released: Supports LE Audio, the next generation of Bluetooth audio. LE Audio provides a noticeable improvement on Bluetooth Classic Audio. It uses less power and less bandwidth while providing high-quality transmissions. An important feature of LE Audio is that it supports the ability to connect to multiple devices at once. Thus, a smartphone, for example, is able to stream audio to two pairs of headsets.

2021: Bluetooth 5.3 released: Primarily supports enhancements to existing features, including *Periodic Advertising* enhancement, which enables at the receiver a more efficient processing of redundant data, resulting in energy savings by the receiver.

2023: Bluetooth 5.4 released: As with 5.3, primarily supports enhancements including *Encrypted Advertising Data* that results in more secure communications.

2024: Bluetooth 6.0 released: Provides key new features such as *Channel Sounding* which lets one determine with high precision the distance and direction between two Bluetooth devices.

Going forward our interest will be in Bluetooth 5 and 6 as these are currently the latest standards available. The capabilities of BLE 5/6 devices over Bluetooth Classic ones are substantial and BLE 5/6 devices are fully backward compatible with older Wi-Fi devices.

1.5   Smartphone GPS Overview

It is not the aim of this text to examine in any detail *Global Positioning System* (GPS) technology as utilized in smartphones. This is because GPS technology is highly complex and has myriad applications of which its use in smartphones is just one. Nonetheless, given that GPS is a wireless technology found in smartphones, a cursory review of its application therein is in order.

GPS is a radio navigation system developed by the U.S. Navy, currently owned by the U.S. government, used to determine locations with approximately 7 m accuracy 95% of the time, anywhere on or close to the earth's surface. It is made up of three parts, namely 31 satellites in medium earth orbit, each orbiting at an altitude of approximately 20,200 km, ground stations, and receivers. The ground station use radar to ensure that the satellites are actually where we think they are. A receiver, such as may be embedded in a smartphone, constantly listens to signals from some of these satellites. It uses the data from these signals to calculate exactly where it is and what time it is with no need to send signals back to the satellites for the system to work.

Each satellite is equipped with its own atomic clock and sends a time coded signal on one of three specific frequencies, namely 1575.42, 1227.6, and 1176.45 MHz. The receiver determines which satellites are visible and gathers data from those with the strongest signals. Through a combination of data received from at least four satellites, the receiver determines location and time.

GPS receivers when computing location use a lot of power. This power consumption can be reduced in smartphones by using Assisted Global Positioning System (AGPS). A smartphone knows approximately where it is, based on its communication with cell towers. With AGPS, the GPS receiver uses this information to more quickly calculate its position thus lowering power consumption.

1.6   Apple's iPhone Satellite Messaging Technology

Apple's iPhone 14 and later models allow one to connect an iPhone via a satellite to request roadside assistance, to text to emergency services, to exchange messages with friends and family, and share location while in areas where there is no cellular or Wi-Fi coverage. With the release of iPhone 14 in 2022, messaging service was only provided for addressing emergencies. However, with the 2024 release of software version iOS 18, the service has been opened for regular, non-emergency two-way messaging for those iPhone users in countries where the service is available (currently only available in the United States and Canada).

As with GPS, it is not the aim of this text to examine in any detail the technology behind Apple's iPhone satellite messaging service. Suffice it to say that this service is provided via the Globalstar satellite communication system. The Globalstar system employs low earth orbit (LEO) satellites, orbiting at an altitude of 1414 km. Figure 1.2 shows a simplified depiction of the system. The Gateway (ground station) shown therein provides connection to the Internet or the Public Switched Telephone Network (PSTN). Outgoing messages from the iPhone are relayed by the satellite to the Gateway which forward them to the receiving smartphone. For incoming messages to the iPhone, the signal transmission flow is reversed.

Globalstar was originally authorized to operate 48 satellites. However, it currently operates 31 and plans to downsize to 26. It operates 24 Gateways spread over six continents, with seven in North America. Coverage is provided ideally from a latitude of about 70° North to a latitude of about 70° South, thus covering over 80% of the Earth's surface but excluding the polar regions. In practice, however, it may not operate above latitude 62° North such as northern parts of Alaska and below latitude 62° South which encompasses all of Antarctica.

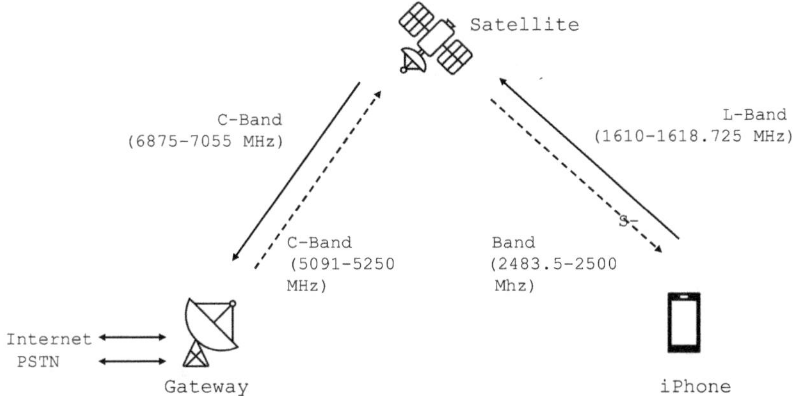

Fig. 1.2 Globalstar/Apple messaging system

1.7 Summary

Smartphones use at least four different wireless communication technologies, namely mobile access, Wi-Fi, Bluetooth, and GPS. Furthermore, in the case of more recent iPhone models, satellite messaging technology is included. In this chapter, brief histories of mobile access, Wi-Fi, and Bluetooth were presented and a cursory look at GPS and iPhone satellite messaging technology was provided.

Chapter 2
Data Communication Systems Protocol Stacks

2.1   Introduction

5G NR, Wi-Fi, and Bluetooth communications are all data based where the data is sent in discrete packets, each packet being a sequence of *bytes* (each byte contains 8 bits). Contained within each packet is its sender's address and the address of the intended recipient. In packet switched networks, the packets at the receiving end do not necessarily arrive in the order sent, so they must be reordered before further use. Here a circuit is configured and left open only long enough to transport an addressed packet, then freed up to allow reconfiguration for transporting the next packet, which may be from a different source and addressed to a different destination. Packet data communication systems employ the so-called protocols. Here a protocol is a set of rules that must be followed for two devices to communicate with each other. Protocol rules are implemented in software and firmware and cover (a) data format and coding, (b) control information and error handling, and (c) speed matching and sequencing. Protocols are layered on top of each other to form a communication architecture referred to as a *protocol stack*. Each protocol provides a function or functions required to make data communication possible. Typically, many protocols are used so that overall functioning can be broken down into manageable pieces.

5G and Wi-Fi networks communicate via the Internet. The Internet communicates with the aid of a suite of software network protocols called TCP/IP or UPD/IP. With TCP/IP and UDP/IP, data to be transferred along with administrative information is structured into a sequence of the so-called *datagrams*, each datagram being a sequence of bytes. TCP/IP and UDP/IP work together with a data link layer and a physical layer (these latter layers combined are referred to as the Network Access Layer) to create an effective packet switched network for Internet communications. The data link layer protocol puts the TCP/IP or UDP/IP datagrams into packets (it is like putting an envelope into another envelope) and the physical layer is responsible for defining procedures for the reliable transmission of data link

D. H. Morais, *5G/5G-Advanced, Wi-Fi 6/7, and Bluetooth 5/6*, https://doi.org/10.1007/978-3-031-82830-0_2

packets between different devices over the physical layer. In the following subsections, a brief introduction to TCP, UDP, IP, data link, and physical layer protocols is presented.

2.2   TCP/IP

The acronym TCP/IP is commonly used in the broad sense and as applied here refers to a hierarchy of protocols occupying five layers but derives its name from the two main protocols, namely TCP and IP. The protocols are stacked as shown in Fig. 2.1. Note that the layers are numbered, with the lowest layer, the physical layer, being labeled Layer 1 and the highest layer, the application layer, being labeled Layer 5. Following is a short description of each protocol layer.

2.2.1   *Application Layer Protocol (Layer 5)*

This is where the user interfaces with the network and includes all the processes that involve user interaction. Protocols at this level key to smartphone TCP/IP communications include the Hyper Text Transfer Protocol (HTTP) which facilitates World-Wide-Web (WWW) access and Simple Mail Transfer Protocol (SMTP) which facilitates the sending of e-mail messages. Data from this level to be sent to a remote address is passed on the next lower layer, the transport layer, in the form of a stream of 8-bit bytes. In the receiving mode, data from the transport layer is passed up to the application layer.

Fig. 2.1 TCP/IP protocol stack

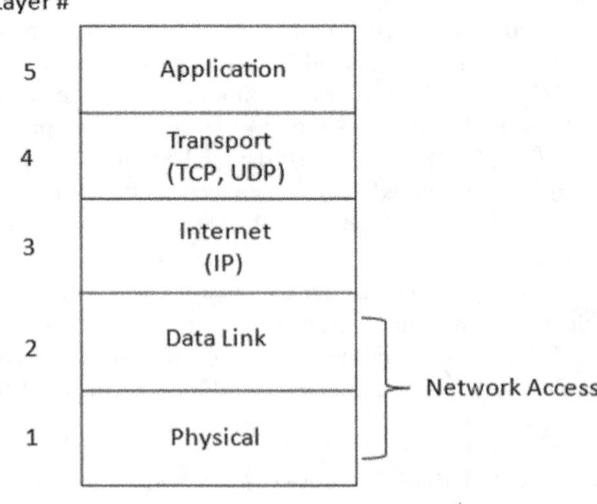

2.2.2   Transport Layer Transmission Control Protocol (Layer 4)

As shown in Fig 2.1, the *Transmission Control Protocol* (TCP) is one of two proto-cols at the Transport layer. TCP provides reliable end-to-end communication and is used by most Internet applications. With TCP, two users, one at each end of the network, must establish a two-way connection before any data can be transferred between them. It is thus referred to as a *connection-oriented* protocol. It guarantees that data sent is not only received at the far end but correctly so. It does this by hav-ing the far end acknowledge receipt of the data. If the sender receives no acknowl-edgment within a specified time frame it resends the data. When TCP receives data from the application layer above for transmission to a remote address, it first splits the data into manageable blocks based on its knowledge of how large a block the network can handle. It then adds control information to the front of each data block. This control information is called a *header* and the addition of the header to the data block is called *encapsulation*. TCP adds a 20-byte header to the front of each data block to form a TCP datagram. It then passes these datagrams to the next layer below, the IP layer. A TCP datagram is shown Fig. 2.2. When TCP receives a data-gram from a remote address via the IP layer below, the opposite procedure to that involved in creating a datagram takes place, i.e., the header is removed, and the data passed to the application layer above. However, before passing it above, TCP takes data that arrives out of sequence and puts it back into the order in which it was sent.

2.2.3   Transport Layer User Datagram Protocol (Layer 4)

The *User Datagram Protocol* (UPD) is the second of two protocols at the Transport layer. Unlike TCP, with UDP, users at each end of the network need not establish a connection before any data can be transferred between them. It is therefore described as a *connectionless* protocol. It provides unreliable service with no guarantee of

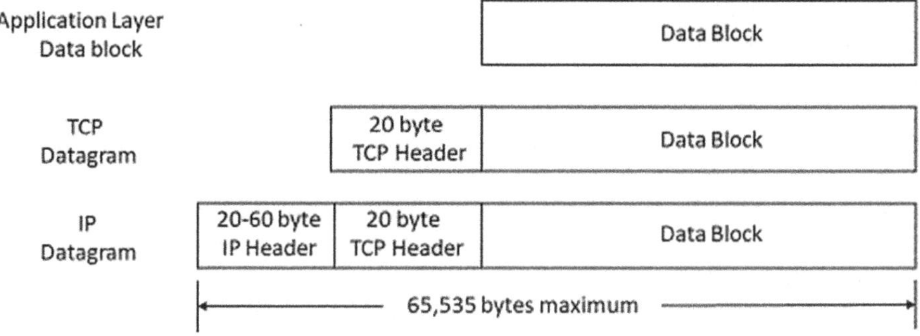

Fig. 2.2 TCP and IP encapsulation

delivery. Packets may be dropped, delayed, or may arrive out of sequence. The service it provides is referred to as a best effort one as all attempts are made to deliver packets, with poor reliability caused only by hardware faults or exhausted resources and the fact that no retransmission is requested in the event of error or loss. UDP adds an 8-byte header to the front of each data block to form a UPD datagram. UPD packets arrive more quickly than TCP ones and are processed faster at the cost of potential error. UPD is used primarily for connections where low latency is important, and some data loss is tolerable. Good examples of such connections are voice and video.

2.2.4 Internet Layer Protocol (Layer 3)

The protocol at this layer, the *Internet Protocol* (IP), is the core protocol of TCP/IP. Its job is to route data across and between networks. When sending datagrams, it figures out how to get them to their destination, when receiving datagrams, it figures out to whom they belong. It is an unreliable protocol, unconcerned as to whether datagrams it sent arrive at their destination and whether datagrams it received arrive in the order sent. If a datagram arrives with any problems, it simply discards it. It leaves the quest for reliable communication, if required, to the TCP level above. Like UDP, it is defined as a *connectionless* protocol. There are, naturally, provisions to create connections per se or communication would be impossible. However, such connections are established on a datagram-by-datagram basis, with no relationship to each other. It processes each datagram as an entity, independent of any other datagram that may have preceded it or may follow it. In fact, there is no information in an IP datagram to identify it as part of a sequence or as belonging to a particular task. Thus, for IP to accomplish its assigned task, each IP datagram must contain complete addressing information. An IP datagram is created by adding a header to the transport layer datagram received from above.

Internet protocols are developed by the *Internet Engineering Task Force* (IETF). *IPv4* was the fourth iteration of IP developed by the IETF but the first version to be widely deployed. It is the dominant network layer protocol on the Internet. Its header is normally 20 bytes long but can in rare circumstances be as large as 60 bytes. A later IP version is IPv6. Its header is 40 bytes long, but 'extension' headers can be added when needed.

An IPv4 datagram created from a TCP datagram is shown in Fig. 2.2. This header includes source and destination address information as well as other control information. As the minimum amount of data transportable is 1 byte, then the minimum size of a data carrying TCP derived IP datagram is 41 bytes. Its maximum size is specified as 65,535 bytes. All IP networks must be able to handle such IP datagrams of at least 576 bytes in length. IP passes datagrams destined to a remote address to the next layer below, the link layer. On the receiving end, datagrams received by the IP layer from the link layer are stripped of their IP headers and passed up to the transport layer. The maximum amount of data that a link

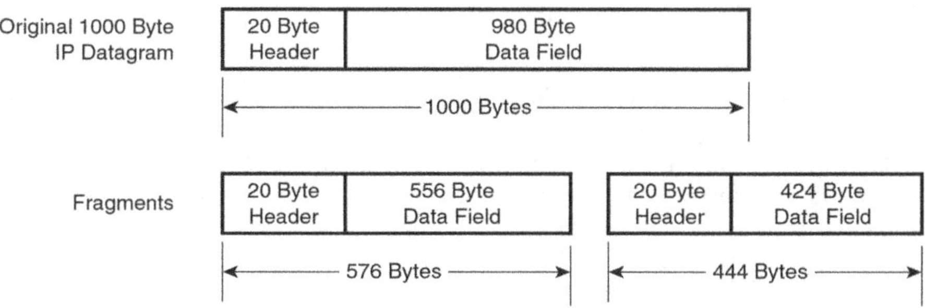

Fig. 2.3 An example of the fragmentation of an IPv4 datagram

layer packet can carry is called the *maximum transfer unit* (MTU) of the layer. Because, as is explained below, each IP datagram is encapsulated within a link layer packet prior to transmission to the remote address, the MTU of the link layer places a maximum on the length of an IP datagram that it can process. Should an IP datagram be larger than this maximum, it is broken up into smaller datagrams that can fit in the link layer packet. This breaking up process is called *fragmentation* and each of the smaller datagrams created is called a *fragment*. Figure 2.3 shows the fragmentation of an IP datagram 1000 bytes long into two fragments, in order that it may be processed by a link layer with an MTU of 576 bytes. It will be noticed that each fragment has the same basic structure as an IP datagram. Information in the fragment header defines it as a fragment, not a datagram. At the destination, fragments are reassembled to the original IP datagram before being passed to the transport layer. Should one or more of the fragments fail to arrive, the datagram is assumed to be lost and nothing is passed to the transport layer. Fragmentation puts additional tasks on Internet hardware and so it is desirable to keep it to a minimum. Obviously, it can be entirely eliminated by using IP datagrams no longer than 576 bytes since all IP handling networks must have an MTU of at least 576.

For an IP datagram created from a UDP datagram carrying data, the minimum data size is 1 byte. As the UDP header is only 8 bytes long and the IPv4 header can be as little as 20 bytes, a UDP generated IPv4 data carrying datagram can be as small as 29 bytes. As with a TCP generated IP datagram, the maximum size of a UDP derived IP datagram is 65,535 bytes.

IPv4 uses 32-bit addresses which limits address space to 4,295 million possible unique addresses. As addresses available are consumed it appears that an IPv4 address shortage is inevitable. Further, IPv4 provides only limited Quality of Service (QoS) capability. It is the limitation in address space and QoS capability of IPv4 that helped stimulate the push to *IPv6* which is the only other standard internet network layer used on the Internet. IPv6 has a 40-byte header and uses 128-bit addresses which results in approximately 300 billion, billion, billion, billion unique addresses! Furthermore, IPv6 provides for true QoS. IPv4 will be supported alongside IPv6 for the foreseeable future.

2.2.5   Data Link Layer Protocol (Layer 2)

This layer creates a "frame" by adding a header and sometimes a trailer to the datagram received from the IP layer above. The header includes, among other information, the source and destination address of the frame and the trailer normally contains bits to aid in error detection. When receiving data, this layer determines if each frame received contains the right host's address and, if a trailer is attached, that the frame is error free. If yes to both, then the frame is stripped of its header and trailer and forwarded upwards to the IP layer.

We note that TCP/IP does not define any specific protocol for the data link layer. This is because there are several different forms of network access. For example, communications over a solid physical link such as that made of twisted pair copper wire or coaxial cable, or optical cable, is typically over *Ethernet* based equipment. Communication over the air for relatively long distances is typically via mobile equipment such as 5G mobile. Communication over the air for relatively short distances is typically via Wi-Fi based equipment. These different forms of network access have their own data link layer protocols which are supported by TCP/IP. The data link layer protocol for 5G NR is described in Sect. 8.5 and that for Wi-FI 6 in Sect. 9.3.1.

Neither 5G NR nor Wi-Fi equipment use the Ethernet protocol as a part of their transmission protocols. However, IP data is normally transported to and from these systems via Ethernet networks. Ethernet frames range in size from 72 to 1526 bytes and Ethernet transmission speeds currently in use vary from 10 MB/s all the way up to 100 Gb/s.

2.2.6   Physical Layer Protocol (Layer 1)

The physical layer is the lowest layer. As with the data link layer, TCP/IP does not define any specific protocol for this layer but supports all of the standard and proprietary protocols. This layer deals with the bit transmission between different devices over the physical medium. This medium may be electrical cable, fiber optic cable, or the air between the devices. It specifies the characteristics of the hardware to be used for a network connection to be established. It stipulates rules for mechanical features such as connector shapes and sizes and rules for electrical features such as voltage, frequency, bit rate, etc.

As with the data link layer, TCP/IP does not define any specific protocol for this layer. The physical layer protocol for 5G NR is described in Sect. 8.6 and that for Wi-Fi 6 in Sect. 9.3.2.

2.3 Bluetooth Protocol Stack

Instead of applying the TCP/IP model, Bluetooth LE has its own independent protocol stack. However, there are many similarities such as the use of packets and frames and a certain equivalence between the Bluetooth LE layers and those of TCP/IP. The Bluetooth LE protocol stack is described in Sect. 10.2.

2.4 Summary

5G and Wi-Fi networks communicate via the Internet. The Internet communicates with the aid of a suite of software network protocols called TCP/IP or UPD/IP. TCP/IP and UDP/IP work together with a data link layer and a physical layer (these latter layers combined are referred to as the network access layer) to create an effective packet switched network for Internet communications. In this chapter, a brief introduction to TCP, UDP, IP, data link, and physical layer protocols was presented.

Chapter 3
The Wireless Path

3.1 Introduction

A smartphone wireless path is between the smartphone and a second terminal. In the case of mobile communications, this second terminal is a base station. In the case of Wi-Fi communications, the second terminal is an access point. Finally, in the case of Bluetooth communications, the second terminal is a device such as a speaker on an in-car entertainment system. Communication over such a path is via the propagation of radio waves. In an ideal world, the path would be free of obstruction, i.e., have a *line-of-sight* (LOS) between transmitter and receiver and attenuate the transmitted signal by a fixed amount across the transmitted signal spectrum, resulting in a predictable and undistorted signal level at the receiver input. Such attenuation is referred to as *free space loss*. In the real smartphone world, however, paths are often not LOS. Rather, they may be non-line of sight (NLOS), there may be multiple paths, and the intervening topography and atmospheric conditions may result in a received signal that deviates significantly from ideal. Changes to the received signal over and above the loss over free space are referred to as fading. In the succeeding sections, we first look at propagation over the smartphone to the mobile base station path. We then look at propagation over the smartphone to Wi-Fi and Bluetooth paths, such paths being of similar characteristics.

3.2 Propagation Over a Mobile Path

A typical base station to mobile unit link is shown in Fig. 3.1. As will be seen in succeeding sections, the maximum length of a wireless path for reliable communications varies depending on (a) the propagation frequency; (b) the antenna heights; (c) terrain conditions between the sites, in particular, the clearance or lack thereof

D. H. Morais, *5G/5G-Advanced, Wi-Fi 6/7, and Bluetooth 5/6*,
https://doi.org/10.1007/978-3-031-82830-0_3

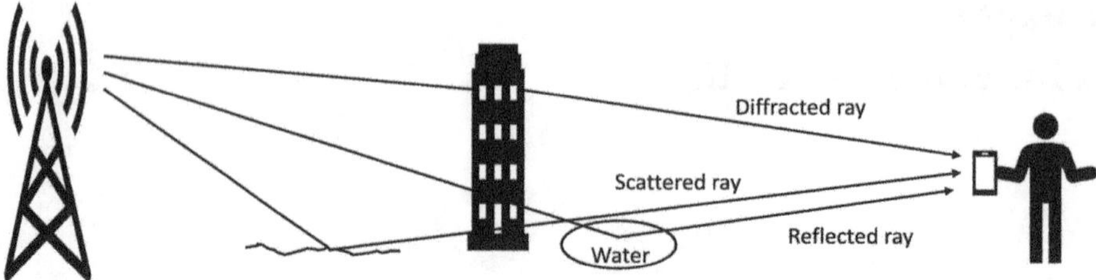

Fig. 3.1 Typical base station to mobile unit link

between the direct signal path and ground obstructions; (d) atmospheric conditions over the path; and (e) the radio equipment and antenna system electrical parameters. For mobile systems operating in the 1–10-GHz bands, paths for reliable communications, depending on the general surrounding topography, can be up to several kilometers in length. For those operating in the 24–100-GHz bands; however, path lengths are much more restricted, being typically less than a kilometer, much of this restriction being due to higher free space loss as a result of the higher frequency and higher obstruction losses. In this chapter, we will examine propagation in an ideal environment and then study the various types of fading and how such fading impacts path reliability. As antennas provide the means to efficiently launch and receive radio waves, a brief review of their characteristics is in order as the first step in addressing ideal propagation.

3.2.1　Antennas

For 5G base stations, antennas provide either omnidirectional coverage, coverage over a sector, typically 120° in azimuth, or highly directional coverage to individual mobile units. For *point-to-multipoint* (PMP) mobile links, the mobile unit antennas are normally omnidirectional in sub-6-GHz bands but can be directional in the millimeter wave bands. A base station omnidirectional antenna is usually a vertical stack of dipole antennas, referred to as a *stacked dipole* or *collinear array*. A base station sector antenna, on the other hand, is usually a linear array of dipole antennas combined electrically so as to produce directivity and is referred to as a flat plane or planar array antenna. Many antenna characteristics are important in designing PMP wireless broadband systems. The most important of these will be reviewed followed by a brief description of those antennas commonly used in such systems.

Antenna gain is the most important antenna characteristic. It is a measure of the antenna's ability to concentrate its energy in a specific direction relative to radiating it isotropically, i.e., equally in all directions. The more concentrated the beam, the higher the gain of the antenna. A directional antenna's gain is function of the square of the frequency; thus, doubling the frequency increases the gain by a factor of 4, i.e., 6 dB. It is also a function of the area of the aperture. Thus, for a parabolic antenna, for example, the gain is a function of the square of the antenna's diameter, and so doubling the diameter increases the gain by a factor of 4, i.e., 6 dB. The operation of an antenna in the receive mode is the inverse of its operation in the transmit mode. It therefore comes as no surprise that its receive gain, defined as the energy received by the antenna compared to that received by an isotropic absorber, is identical to its transmit gain. Figure 3.2 shows a plot of antenna gain versus angular deviation from its direct axis, which is referred to as its *pole* or *boresight*. It shows the main lobe where most of the power is concentrated. It also shows side lobes and back lobes that can cause interference into or from other wireless systems in the vicinity.

The *beamwidth* of an antenna is closely associated with its gain. The higher the gain of the antenna, the narrower the width of the beam. Beamwidth is measured in radians or degrees and is usually defined as the angle that subtends the points at which the peak field power is reduced by 3 dB. Figure 3.2 illustrates this definition. The narrower the beamwidth, the more interference from external sources including nearby antennas is minimized. This improvement comes at a price, however, as narrow beamwidth antennas require precise alignment. This is because a very small shift in alignment results in a measurable decrease in the transmitted power directed at a receiving antenna and the received power from a distant transmitting antenna.

A radiated electromagnetic wave consists of an electrical field and a magnetic field, each advancing in a plane at 90 degrees to the other. The polarization of an antenna refers to the alignment of the electric field in the radiated wave. Thus, in a *horizontally polarized* antenna, the electric field is horizontal, and in a *vertically polarized* antenna, the electric field is vertical.

Finally, we note that an antenna related parameter often regulated for wireless systems is the *Equivalent Isotropically Radiated Power* (*EIRP*). This power is the product of the power supplied to the transmitting antenna, P_t, say, and the gain of the transmitting antenna, G_{ta} say. Thus, we have

$$EIRP = P_t \cdot G_{ta} \tag{3.1}$$

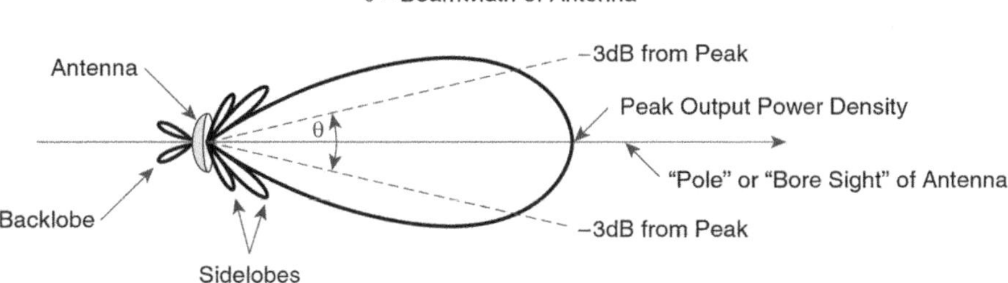

Fig. 3.2 Antenna gain versus angular deviation from its pole

3.2.2   Free Space Loss

As indicated above, the propagation of a signal over a wireless path is affected by both atmospheric anomalies and the intervening terrain. Absent of any such interfering effects, we have signal loss only as a result of free space. Free space loss is defined as the loss between two isotropic antennas in free space, an isotropic antenna being one that radiates equally in all directions. It can be shown [1] that the free space loss, L_{fs}, experienced between two isotropic antennas is a function of frequency and distance and is given by

$$L_{fs}\left(\text{dB}\right) = 32.4 + 20\log_{10} f + 20\log_{10} d \tag{3.2}$$

where f is the transmission frequency in MHz; d is the transmission distance in km.

To get a sense of the magnitude of free space loss typical of wireless broadband links, consider a 2-GHz path, 3 km long, and a 25-GHz path 0.3 km long. For the 2-GHz path, the free space loss would be 108 dB, and for the 25-GHz path, the loss would be 110 dB.

Thus, even at one-tenth the distance of the 2-GHz path, the 25-GHz path suffers slightly more loss than the 2 GHz path.

3.2.3   Line-of-Sight Non-Faded Received Signal Level

With a relationship to determine free space loss, one is now able to also set out a relationship for the determination of the receiver input power P_r in a typical wireless link assuming no loss to fading or obstruction. Starting with the transmitter output power P_t, one simply accounts for all the gains and losses between the transmitter output and the receiver input. These gains and losses, in dBs, in the order incurred are:

L_{tf} = loss in transmitter antenna feeder line if such a line exists (coaxial cable or waveguide depending on frequency)

G_{ta} = gain of the transmitter antenna

L_{fs} = free space loss

G_{ra} = gain of the receiver antenna

L_{rf} = loss in the receiver antenna feeder line if such a line exists

Thus

$$P_r = P_t - L_{tf} + G_{ta} - L_{fs} + G_{ra} - L_{rf} \tag{3.3}$$

Example 3.1

Computation of received input power

A PMP link has the following typical parameters:

Path length, $d = 3$ km

Operating frequency, $f = 2$ GHz

Transmitter output power, $P_t = 40.0$ dBm (10 W)
Loss in transmitter antenna feeder line, $L_{tf} = 2$ dB
Transmitter antenna gain, $G_{ta} = 10$ dB
Receiver antenna gain, $G_{ra} = 2$ dB
Loss in receiver antenna feeder line, $L_{rf} = 0.2$ dB

What is the receiver input power?

Solution

By Eq. (3.2), the free space loss, L_{fs}, is given by

$$L_{fs} = 32.6 + 20\log_{10} 2,000 + 20\log_{10} 3 = 108.2 \text{dB}$$

Thus, by Eq. (3.3), the receiver input power, P_r, is given by

$$P_r = 40 - 2 + 10 - 108.2 + 2 - 0.2 = -58.4 \text{ dBm}$$

For a LOS wireless link, received input power P_r resulting from free space loss only is normally designed to be higher than the minimum receivable or threshold level R_{th} for acceptable probability of error. This power difference is engineered so that if the signal fades, due to various atmospheric and terrain effects, it will fall below its threshold level for only a small fraction of time. This built-in level difference is called the link *fade margin* and thus given by

$$\text{Fade margin} = P_r - R_{th} \tag{3.4}$$

3.2.4   *Fading Phenomena*

The propagation of radio signals through the atmosphere is affected by terrain features in or close to its path in addition to the atmospheric effects. Potential propagation affecting terrain features are trees, hills, sharp points of projection, reflection surfaces such as ponds and lakes, and man-made structures such as buildings and towers. These features can result in either reflected or diffracted signals arriving at the receiving antenna. Potential propagation affecting atmospheric effects, on relatively short paths, includes rain, water vapor, and oxygen.

3.2.4.1   Fresnel Zones

The proximity of terrain features to the direct path of a radio signal impacts their effect on the composite received signal. *Fresnel zones* are a way of defining such proximity in a very meaningful way. In the study of the diffraction of radio signals, the concept of Fresnel zones is very helpful. Thus, before reviewing diffraction, an

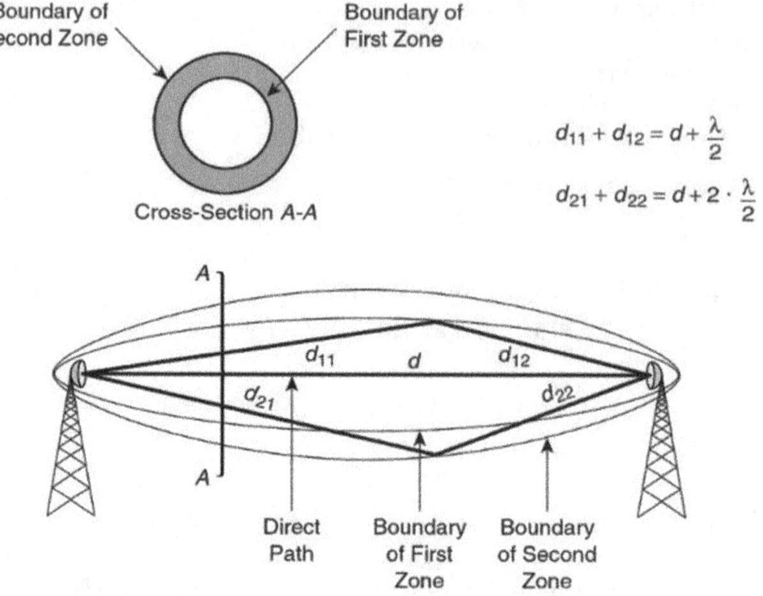

Fig. 3.3 First and second Fresnel zone boundaries

short overview of Fresnel zones is in order. The first Fresnel zone is defined as that region containing all points from which a wave could be reflected such that the total length of the two-segment reflected path exceeds that of the direct path by half a wavelength, $\lambda/2$, or less. The nth Fresnel zone is defined as that region containing all points from which a wave could be reflected such that the length of the two-segment reflected path exceeds that of the direct path by more than $(n - 1)\lambda/2$ but less than or equal to $n\lambda/2$. Figure 3.3 shows the first and second Fresnel zone boundaries on a LOS path. The perpendicular distance, F_1, from the direct path to the outer boundary of the 1st Fresnel zone is approximated by the following equation:

$$F_1 = 17.3 \left[\frac{d_{11}d_{12}}{fd} \right]^{\frac{1}{2}} \text{m} \qquad (3.5)$$

where d_{11}, d_{12}, and d are in km and f is in GHz.

Thus, for a 1-km path operating at 2 GHz, the maximum value of F_1, which occurs at the middle of the path, is, by Eq. (3.5), 6.1 m. However, for a typical 1 km, 25-GHz path, F_1 maximum is only 1.73 m.

Most of the power that reaches the receiver is contained within the boundary of the first Fresnel zone (more on this in the next section). Thus, terrain features that lie outside this boundary, with the exception of highly reflective surfaces, do not, in general, significantly affect the level of the received signal. Depending on how much phase shift is added to a reflected signal at this boundary, the sum of the direct signal and the reflected signal at the receiving antennas can be additive, resulting in a composite signal that is as much as 6 dB greater than the direct signal. It can also be subtractive, however, resulting in theory in a signal infinitely small.

3.2.4.2   Reflection

The reflection of a radio signal occurs when that signal impinges on a surface which has very large dimensions compared to the wavelength of the signal. Reflections typically occur from the surface of the earth, particularly liquid surfaces, and from buildings. When reflection occurs, the reflected signal arriving at a receiving antenna is combined with the direct signal to form a composite signal that, depending on the strength and phase of the reflected signal, can significantly degrade the desired signal. Figure 3.4 shows a radio path with a direct signal and a reflected signal off a lake.

The strength and phase of a reflected signal relative to the incident signal at the point of reflection are strongly influenced by the composition of the reflecting surface, the curvature of the surface, the amount of surface roughness, the incident angle, and the radio signal frequency and polarization. If reflections occur on rough surfaces, they usually do not create a problem as the incident and reflected angles are quite random. Reflections from a relatively smooth surface, however, depending on the location of the surface, can result in the reflected signal being intercepted by the receiving antenna.

3.2.4.3   Diffraction

So far in explaining propagation effects, it has been tacitly assumed that the energy received by a radio antenna travels as a beam from transmitting to receiving antenna that's just wide enough to illuminate the receiving antenna. This is not exactly the case, however, as waves propagate following the *Huygens' principle*. Huygens showed that propagation occurs along a wavefront, with each point on the wavefront acting as a source of a secondary wavefront known as a wavelet, with a new wavefront being created from the combination of the contributions of all the wavelets on the preceding front. Importantly, secondary wavelets radiate in all directions. However, they radiate strongest in the direction of the wavefront propagation and less and less as the angle of

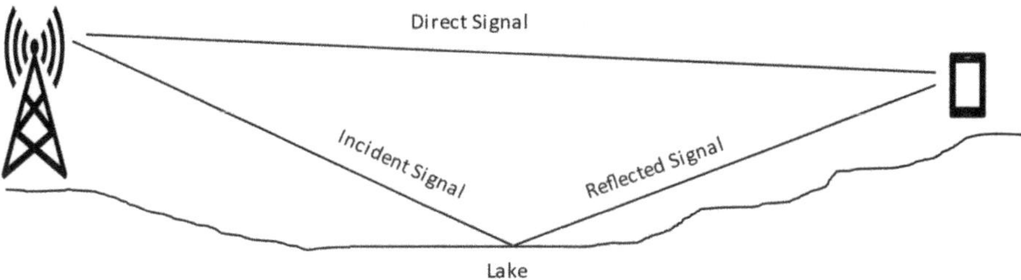

Fig. 3.4 Reflected signal

radiation relative to the direction of the wavefront propagation increases until the level of radiation is zero in the reverse direction of wavefront propagation. The net result is that, as the wavefront moves forward, it spreads out, albeit with less and less energy on a given point as that point is removed further and further from the direct line of propagation. This sounds like bad news. The signal intercepted at the receiving antenna, however, is the sum of all signals directed at it from all the wavelets created on all wavefronts as the wave moves from transmitter to receiver. At the receiver, signal energy from some wavelets tend to cancel signal energy from others depending on the phase differences of the received signals, these differences being generated as a result of the different path lengths. The net result is that, in a free space, unobstructed environment, half of the energy reaching the receiving antenna is canceled out.

Signal components that have a path difference of $\lambda/4$ or less relative to the direct line signal are additive. Signals with a path difference between $\lambda/4$ and $\lambda/2$ are subtractive. In an unobstructed environment, all such signals fall within the first Fresnel zone. In fact, the first Fresnel contains most of the power that reaches the receiver. Consider now what happens when an obstacle exists in a radio path within the first Fresnel zone. Clearly, under this condition, the amount of energy intercepted at the receiving antenna will differ from that intercepted if no obstacle were present. The cause of this difference, which is the disruption of the wavefront at the obstruction, is called *diffraction*. If an obstruction is raised in front of the wave so that a direct path is just maintained, the power reaching the receiver will be reduced, whereas the simplistic narrow beam model would suggest that full received signal would be maintained. A positive aspect of diffraction is that if the obstruction is further raised, so that it blocks the direct path, signal will still be intercepted at the receiving antenna, albeit at a lesser and lesser level as the height of the obstruction increases further and further. Under the simplistic narrow beam model, one would have expected complete signal loss.

Figure 3.5a shows "unobstructed" free space propagation, where path clearance is assumed to exceed several Fresnel zones. Figure 3.5b shows diffraction around an obstacle assumed to be within the region of the first Fresnel zone but not blocking the direct path. Figure 3.5c shows diffraction around an obstacle blocking the direct path. For simplicity an expanding wave that has progressed partially down the path is shown in all three depictions, to the point of obstruction in the case of the obstructed paths. All wavelets on the wavefront shown in the unobstructed case are able to radiate a signal that falls on the receiving antenna. In the case of the obstructed paths, however, the size of the wavefronts and hence the number of wavelets radiating signals that fall on the receiving antenna are decreased. Much research and analysis of diffraction loss have been conducted over several decades leading to well-accepted formulae [2] for its estimation.

3.2.4.4   Scattering

Scattering occurs when a radio signal hits a large rough surface or objects with dimensions that are small compared to signal wavelength. The energy reflected from such surfaces tends to be spread out (scattered) in many directions, and

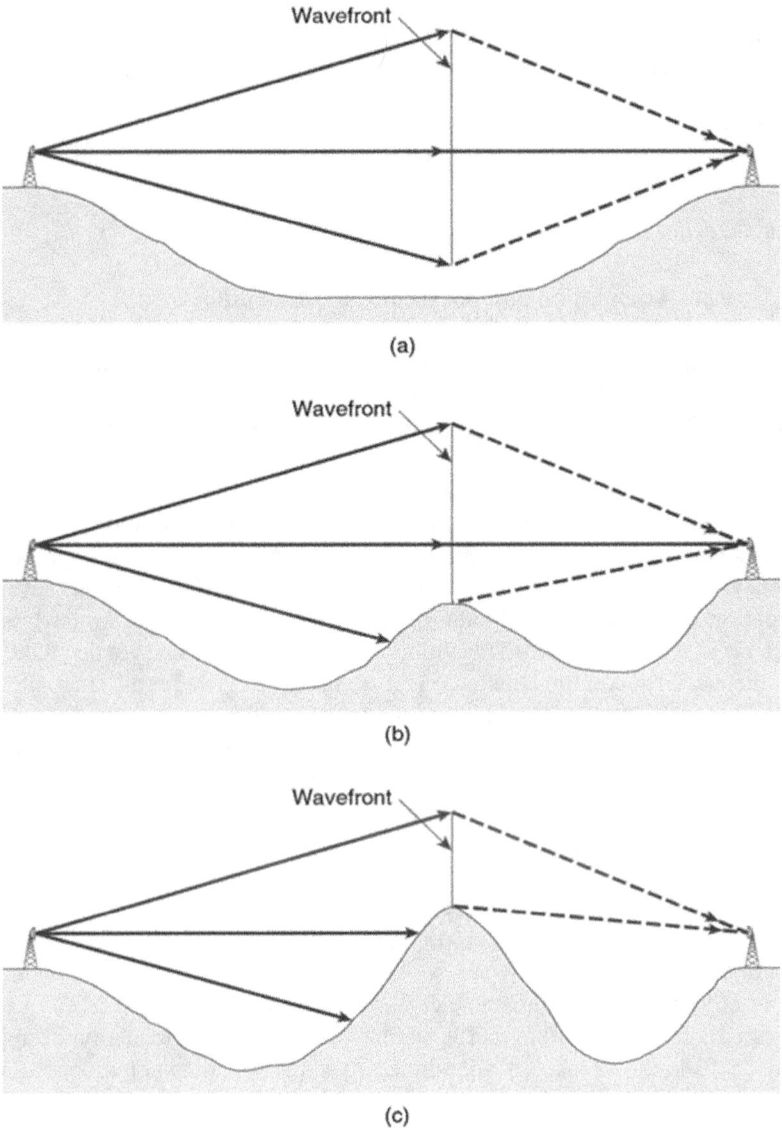

Fig. 3.5 Graphical representation of diffraction. (**a**) Unobstructed path. (**b**) Path with obstacle that does not block direct path. (**c**) Path with obstacle that blocks direct path

thus some of this energy may impinge on the receiver. Figure 3.6 shows a scattering scenario. Typical urban scatterers include rough ground, foliage, street signs, and lamp posts. Often, in a mobile radio environment, the received signal is stronger than that which would have been expected as a result of reflection and diffraction due to the added energy received via scattering. Of course, the phase of the received scattered signal could be such as to reduce overall signal strength.

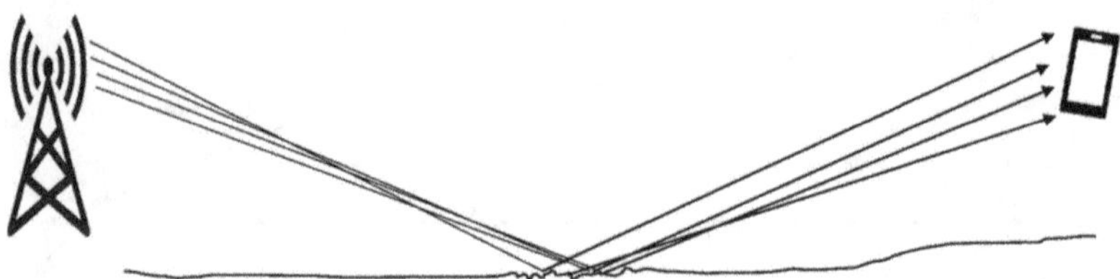

Fig. 3.6 Scattering scenario

3.2.4.5   Rain Attenuation and Atmospheric Absorption

As a radio signal propagates down its path, it may find itself subjected to, in addition to the possible effects of reflection, diffraction, and scattering, the attenuating effects of rain and absorption by atmospheric gasses, primarily water vapor and oxygen.

Raindrops attenuate a radio signal by absorbing and scattering the radio energy, with this effect becoming more and more significant as the wavelength of the signal decreases toward the size of the raindrops. Attenuation increases with propagation frequency and rain intensity up to a frequency of about 100 GHz than levels off. Attenuation due to a heavy cloudburst at 6 GHz is about 1 dB/km, whereas at 24 GHz, it is about 20 dB/km. The width of cloudburst cells tends to be on the order of kilometers, so maximum attenuation due to rain at 24 GHz for a 1 km. path would be approximately 20 dB. In general, rain attenuation is not a major problem below about 10 GHz. However, this situation changes measurably as the frequency increases and normally must be factored in at frequencies of 10 GHz and above.

Attenuation due to water vapor has a first peak at approximately 22 GHz with a value of about 0.18 dB/km. It then declines as the frequency increases up to about 31 GHz, after which it increases back to about 0.18 dB/km in the region of 60 GHz.

Attenuation due to dry air is below 0.01 dB/km for frequencies up to 20 GHz and then increases somewhat exponentially to a peak of approximately 15 dB/km in the region of 60 GHz.

Paths close to 60 GHz, unfortunately, have the worst of two worlds. Rain attenuation can approach 40 dB/km and is additive to the dry air attenuation of approximately 15 dB/km. As a result, PMP wireless systems operating at frequencies in this region and in locations subject to any appreciable rainfall are usually limited in path length to much less than 1 km.

3.2.4.6   Penetration Loss

When a mobile unit is indoors, the received signal suffers additional loss over and above that suffered as a result of outdoor effects as a result of the signal having to penetrate either a wall, or a window, or a door, or a combination thereof. This added loss is thus referred to as *penetration loss* and can be quite large.

Different materials commonly used in building construction have very diverse penetration loss characteristics. Common glass tends to be relatively transparent with a rather weak increase of loss with higher frequency due to conductivity losses. Energy efficient glass commonly used in modern buildings or when renovating older buildings is typically metal-coated for better thermal insulation. This coating introduces additional losses that can be as high as 40 dB even at lower frequencies. Materials such as concrete or brick have losses that increase rapidly with frequency. [3].

Typical building exteriors are constructed of many different materials, e.g., concrete, sheet rock, wood, brick, and glass. Thus, depending on the location of the receiver within the building, its final signal may be a combination of several signals traversing several different materials with different penetration loss properties.

3.2.5 *Signal Strength Versus Frequency Effects*

3.2.5.1 Introduction

Fading in PMP systems is the variation in strength of a received radio signal due normally to terrain and atmospheric effects in the radio's path as discussed in Sect. 3.2.4 above. Fading is normally broken down into two main types based on the impact of fading on the signal spectrum. If fading attenuates a signal uniformly across its frequency band, fading is referred to as *flat fading*. Fading due to rain or atmospheric gasses is typically flat fading. If fading results in varying attenuation across the signal's frequency band, such fading is called *frequency selective fading*. These two types of fading can occur separately or together. It cannot be predicted with any accuracy exactly when fading is likely to occur.

3.2.5.2 Frequency Selective Fading

If a radio path passes over highly reflective ground or water or a building surface and the antenna heights allow, in addition to a direct path, a reflected path between the antennas, reflection-induced fading is likely and, under certain circumstances, can be substantial. Reflection-induced fading is frequency selective. The reflected signal will have, at any instant, a path length difference and hence a time delay, τ say, relative to the direct path that's independent of frequency. However, this delay, measured as a phase angle, varies with frequency. Thus, if the propagated signal occupies a band between frequencies f_1 and f_2 say, then the relative phase delay in radians at frequency f_1 will be $2\pi f_1 \tau$, the relative phase delay at f_2 will be $2\pi f_2 \tau$, and the difference in relative phase delay between the component of the reflected signal at f_1 and that at f_2 will be $2\pi(f_2 - f_1)\tau$. Because the reflected signal will have a phase shift relative to the direct signal that is a function of frequency, then, when both these signals are combined, a different resultant signal relative to the direct signal will be created for each frequency. Depending on the value of τ and the signal bandwidth $f_2 - f_1$, significant differences can exist in the composite received signal amplitude and phase as a function of frequency as compared to the undistorted direct signal. Furthermore, the closer the reflected signal is in amplitude to the direct signal, the larger the maximum signal cancelation possible, this occurring when the signals are 180° out of phase.

3.2.6   NLOS Path Analysis Parameters

For point-to-multipoint NLOS paths, path analysis consists of estimating, for a given distance d from the base station, the following parameters:

- *Mean path loss* of all possible paths of length d.
- *Shadowing*, i.e., variation about this mean due to the impact of varying surrounding terrain features from one path to the next.
- *Multipath fading*, i.e., distortion and/or amplitude variation of the composite received signal resulting from the varying delay in time between the various received signals. These varying time delays result from movement in the surrounding environment, for example, moving vehicles, people walking, and swaying foliage.
- *Doppler shift fading*, i.e., distortion and/or amplitude variation of the composite received signal resulting from variations in the frequencies of received signal components as a result of motion.

3.2.6.1   Mean Path Loss

As indicated earlier, for LOS paths with good clearance, signal loss between a transmitter and receiver antenna is free space loss, given by

$$L_{fs} = 32.4 + 20\log f + 10\log d^2 \qquad (3.6)$$

where f is in MHz and d is in km, and loss in dBs.

The equation above indicates that the loss is proportional to the square of distance. For NLOS (shadowed) paths, however, and not surprisingly, this relationship does not hold, since the composite received signal is sum of a number of signals attenuated by diffraction, reflection, scattering, and atmospheric effects, if any. Rather, it has been found empirically that in such situations, the ensemble average loss $\overline{L_p}(d)$ (in dBs) of all possible paths on circle of radius d from the base station can be stated as follows:

$$\overline{L_p}(d) = L_p(d_0) + 10\log\left(\frac{d}{d_0}\right)^n \qquad (3.7)$$

where d_0 is a close-in reference distance around base station free of obstruction; $L_p(d_0)$ is loss at d_0 and is either measured or computed as free space loss; n is the *path loss exponent*.

For mobile access cells, d_0 is often specified as 100 m for macro-cells and as low as 1 m for microcells. Path loss exponent n can vary from 2 in free space up to about 5 in heavily shadowed urban environments. Note that in a shadowed environment, doubling the path distance increases path loss by $3n$ dB. Thus, for $n = 2$ (free space), loss increases by 6 dB, but for $n = 5$ say, loss increases by 15 dB! Models have been devised that estimate n in terms of terrain conditions for a specific frequency and specific base station and remote station antenna heights. To use these models for other frequencies and other antenna, correction terms are added to the loss deter-mined by applying n in the basic path loss equation. Since $\overline{L_p}(d)$ helps character-ize signal strength over "large distances" (a change in d of a few wavelengths will not change $\overline{L_p}(d)$ in a measurable way), $\overline{L_p}(d)$ is referred to as a *large-scale loss* component.

3.2.6.2   Shadowing

As indicated in the preceding, $\overline{L_p}(d)$ is the ensemble average loss for all paths on the circle of distance d from the base station. The actual loss at each location, how-ever, can be significantly different from this average as the surrounding terrain fea-tures can vary greatly from location to location. The distribution of measured loss about the average loss for a large number of measurements made at the same base station to terminal distance is referred to as *shadowing*. Figure 3.7 shows a simpli-fied presentation of path loss versus distance for a mobile cell. The ensemble aver-age $\overline{L_p}(d)$ is shown as well as boundaries around it wherein most but not all of measured received signals are likely to lie.

3.2.6.3   Multipath Fading

At a receiver in a NLOS path, the received signal typically consists of multiple ver-sions of transmitted signal as a result of reflection, diffraction, and scattering. These signals all arrive at receiver at slightly different times and from different directions, given that they traverse different paths. As a result, they have different amplitudes and phases and add up vectorially at receiver to form a composite received signal. A channel with such multiple signals is called a multipath channel. Figure 3.8 illus-trates such a channel.

If the amplitudes and phases of the received signals vary due to movement of the remote unit, then the composite signal's amplitude and phase also vary. These signal variations are referred to as *multipath fading*. In a NLOS path with a stationary remote unit, the received signal may still suffer multipath fading due to movement in the surrounding environment, for example, moving vehicles, people walking, and swaying foliage. Even the smallest movements cause variations in amplitudes and

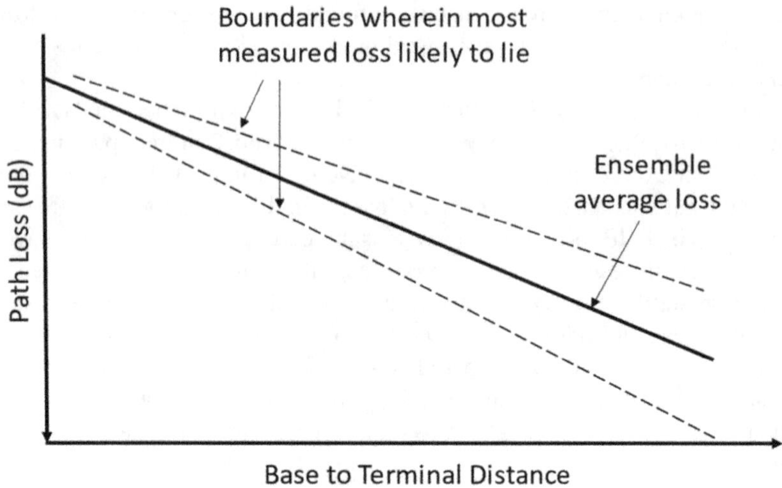

Fig. 3.7 Path loss versus distance for a typical mobile cell

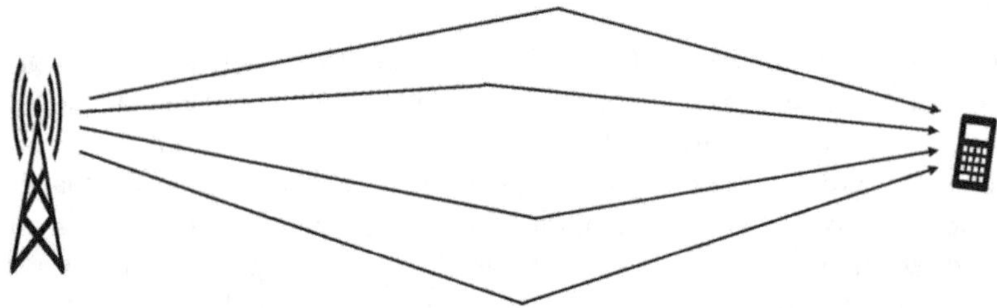

Fig. 3.8 Multipath channel

phases of reflected signals, and the rapidity of these movements translates into rapidity of induced fading. Such rapid fading is therefore referred to as *small-scale fading*. Figure 3.9a depicts small-scale fading as a function of time at a fixed remote unit, and Fig. 3.9b depicts small-scale fading as a function of displacement from the base station at a mobile remote unit.

NLOS multipath fading can be flat or frequency selective. As indicated above, fading is considered flat if the channel has a relatively constant gain and linear phase over the bandwidth of the transmitted signal; whereas, it is considered frequency selective if the channel has significantly varying gain and phase over the bandwidth of the transmitted signal. Figure 3.10 shows signal spectral density versus frequency. In Fig. 3.10a, the transmission channel experiences flat fading as the spectral density across a spectrum much broader than the transmission channel is fairly smooth; thus, even though the location of peaks and valleys are likely to change with time, the variation across the transmission channel is likely to remain fairly flat. In Fig. 3.10b, the peaks and valleys of the broader spectrum are quite sharp. Thus, though the transmission channel is shown experiencing flat fading, it is

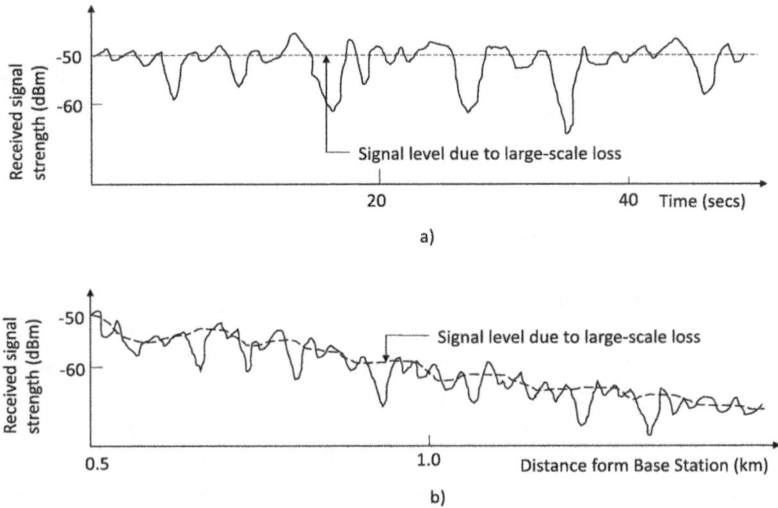

Fig. 3.9 Small-scale fading superimposed on large-scale loss. (**a**) Received signal strength at a fixed remote unit showing small-scale fading superimposed on received signal level due to large-scale loss. (**b**) Received signal strength at mobile remote unit showing small-scale fading superimposed on received signal level due to large-scale loss

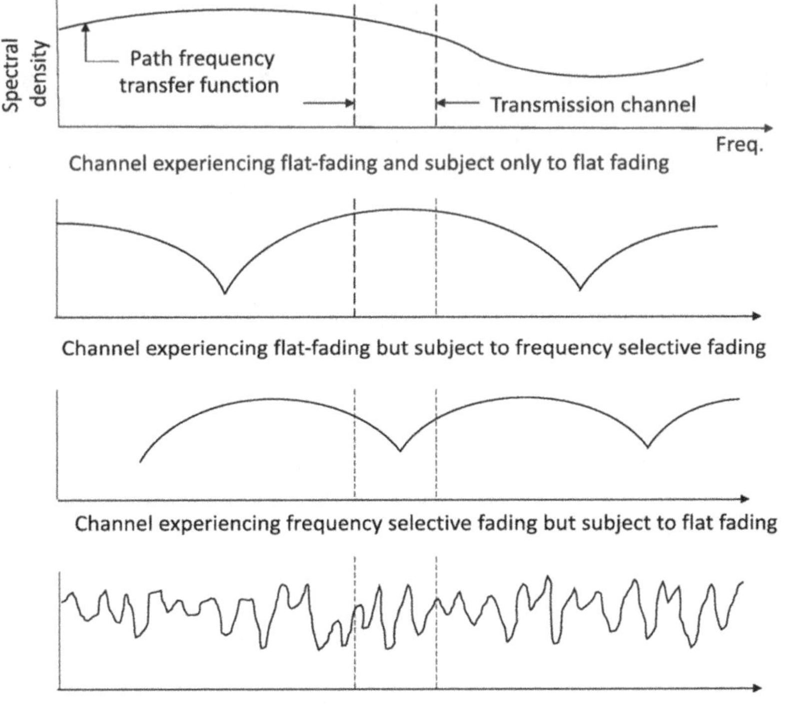

Fig. 3.10 Channels experiencing flat fading and frequency selective fading

subject to frequency selective fading as the broader spectrum shifts with time. In Fig. 3.10c, the situation is the reverse of that in Fig. 3.10b. The transmission channel is shown experiencing frequency selective fading but subject to flat fading. Figure 3.10d shows a transmission channel subject only to frequency selective fading.

3.2.6.4   Doppler Shift Fading

In addition to amplitude/phase signal changes caused by changes in path geometries, a received composite signal is also affected by motion in the surrounding environment. Each signal received by a moving object or each multipath wave that's reflected off a moving object experiences an apparent shift in frequency, this shift being called a *Doppler shift*. This shift is proportional to the speed with which the object moves and a function of the direction of motion. The result is that the spectrum of the received signal is broadened. The effect of this broadening is again fading, in the form of signal distortion, amplitude fluctuations, or a mix of both. Like multipath fading, Doppler shift fading is rapid hence it being classified as small-scale fading. Figure 3.11 demonstrates the impact of speed and direction on apparent wavelength.

3.2.6.5   Millimeter Wave Communications

Broadly stated, one of the goals of 5G NR is to operate in frequencies from below 1 GHz to as high as 100 GHz. In Rel. 15 and early Rel. 16, however, NR supports operation in two specified frequency ranges:

FR1: In 3GPP TS 38.104 Rel. 15, ver. 15.1.0, bands starting as low as 450 MHz and ending as high as 6 GHz, referred to as the sub-6 GHz range. In ver. 15.5.0, however, upper end of range extended to 7.125 GHz. Wider range unchanged in early Rel. 16 versions.
FR2: In Rel. 15 and early Rel. 16 versions, bands starting as low as 24.25 GHz and ending as high as 52.60 GHz, commonly referred to as the millimeter wave (mmwave) range.

Since both the operating frequencies and the channel bandwidths of NR millimeter wave signals are much higher than those in FR1, it is not surprising that propagation analysis for millimeter wave bands require more than simply scaling those developed for the FR1.

Fig. 3.11 Impact of speed on apparent wavelength

As indicated in Sect. 3.2.4.5 above, attenuation due to dry air, water vapor, and rain all increase rapidly above 10–20 GHz.

As with atmospheric effects, diffraction loss is measurably higher at millimeter wave frequencies compared to sub-6-GHz frequencies. The reason for this is readily apparent if we consider the respective first Fresnel zone boundaries. Such boundaries are inversely proportional to the square root of the frequency. The first Fresnel zone boundary of a 25-GHz signal is thus 3.2 times smaller than that of a 2.5-GHz signal assuming the same path length. Since most of the power that reaches the receiver is contained within the boundary of the first Fresnel zone, then power from a 25-GHz signal will be more quickly blocked by an obstruction than power from a 2.5 GHz one. Given the greater diffraction loss at millimeter wave frequencies compared to sub-6-GHz ones, coupled with higher free space loss, in many situations LOS propagation at millimeter wave frequencies may offer acceptable performance where NLOS propagation may not.

Like diffraction loss, penetration loss also tends to increase with frequency. At 28 GHz, penetration loss was measured by comparing path loss outside by the window and indoors 1.5 m from the window [4]. Median loss was 9 dB for plain glass windows and 15 dB for low-emissivity (low-e) windows. Measurements at 38 GHz [5] found a penetration loss of nearly 25 dB for a tinted glass window and 37 dB for a glass door.

Millimeter wave propagation analysis in a mobile environment is clearly a challenging subject, and additional work needs to be done to achieve results as reliable as those achieved for sub-6-GHz propagation. However, many organizations are addressing this subject, and at some time in the future, competing models will no doubt converge to an acceptable form.

3.2.6.6 Path Loss Models

A number of path loss models applicable to 5G mobile systems have been introduced by major organizations or groups thereof in the recent past. They all typically provide both LOS and NLOS versions, and input parameters typically include:

- A general description of the surrounding topography, e.g., indoor, dense urban, urban, and rural
- Base station antenna height
- Mobile unit height
- Ground-level distance between base station and mobile unit
- Direct distance between the base station antenna and mobile unit antenna
- An assumed shadow fading standard deviation in dBs
- Transmission frequency

Among these new models are:

1. The ITU model: ITU-R M.2412-0 [6]
2. The 3GPP model: 3GPP TR 38.901 [7]

3. Model by an ad hoc group of companies and universities (5GCM) [8]
4. The European Union sponsored METIS model [9]
5. The European Union sponsored Millimeter Wave Based Mobile Radio Access Network for Fifth Generation Integrated Communications (mmMAGIC) model [10]

3.3   Propagation Over Wi-Fi and Bluetooth Paths

A significant difference between typical mobile access paths and Wi-Fi or Bluetooth paths is path length. For mobile paths, lengths can vary typically from a few hundred meters when operating in millimeter wave bands to several kilometers when operating in FR1 range bands. For Wi-Fi paths, lengths can be up to about 45 m when operating indoor in the 2.4-GHz band and up to about 15 m when operating indoor in the 5-GHz band. For Bluetooth, which operates only in the 2.4-GHz band, path lengths can be up to about 240 m for very low-rate transmission. Another difference is the likelihood of much more obstruction in Wi-Fi and Bluetooth paths relative to mobile access paths.

With Wi-Fi and Bluetooth paths propagation is typically via LOS signals, and or a combination of reflected signals, diffracted signals, and signals through walls and windows. Given the short distances involved atmospheric effects are not a factor and relative motion is slow enough that Doppler effects are also not a factor. Figure 3.12 depicts propagation paths between a smartphone and a Wi-Fi access point and a Bluetooth speaker.

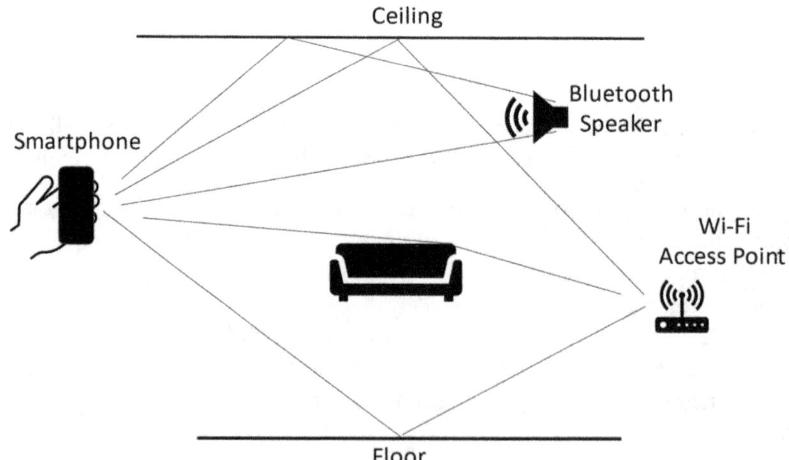

Fig. 3.12 Smartphone/Wi-Fi access point communication paths and smartphone/Bluetooth speaker communication paths

Most Wi-Fi and Bluetooth communication is typically indoors. Propagation prediction for indoor radio transmission differs somewhat from that for outdoor transmission. To help address the need for indoor propagation prediction, the ITU has published Recommendation ITU-R P.1238-11 [11].

3.4 Summary

A smartphone wireless path is between the smartphone and a second terminal. In the case of mobile communications, this second terminal is a base station. In the case of Wi-Fi communications, the second terminal is an access point. Finally, in the case of Bluetooth communications, the second terminal is a device such as a speaker on an in-car entertainment system. In this chapter, we looked at propagation between (a) the smartphone and the mobile base and (b) the smartphone and the Wi-Fi access point/the Bluetooth terminal.

References

1. Morais DH (2022) Key 5G physical layer technologies, 2nd edn. Springer, Cham
2. ITU Recommendation ITU-R P.526–7 (2001) Propagation by diffraction. ITU, Geneva
3. Haneda K et al (2016) 5G 3GPP-like channel models for outdoor urban microcellular and macrocellular environments. In: 2016 IEEE international conference on communications workshops (ICCW), May 2016
4. Holma H et al (eds) (2020) 5G technology; 3GPP new radio. Wiley, Hoboken
5. Rodriguez I et al (2015) Analysis of 38 GHz mmwave propagation characteristics of urban scenarios. European wireless 2015, proceedings of 21st European wireless conference, pp 1–8, May 2015
6. ITU Recommendation ITU-R PM.2412-0 (2017) Guidelines for evaluation of radio interface technologies for IMT-2020. ITU, Geneva
7. 3GPP (2018) 5G; Study on channel model for frequencies from 0.5 to 100 GHz. 3rd Generation Partnership Project (3GPP), TR 38.901 V 14.3.0, Jan 2018
8. 5GCM (2016) 5G channel model for bands up to 100 GHz. Technical Report, Oct 2016
9. METIS (2020) METIS Channel Model. Technical Report METIS2020, Deliverable D1.4 v3, July 2015
10. mmMAGIC (2017) Measurement results and final mmmagic channel models. Technical Report H2020-ICT-671650-mmMAGIC/D2.2, May 2017
11. ITU Recommendation ITU-R P.1238-11 (2021) Propagation data and prediction methods for the planning of indoor radiocommunication systems and radio local area networks in the frequency range 300 MHz to 450 GHz. ITU, Geneva

Chapter 4
Digital Modulation: The Basic Principles

4.1 Introduction

The fundamental modulation methods used in 5G/5G-Advanced are Pi/2 BPSK, QPSK, 16-QAM, 64-QAM, 256-QAM, and 1024-QAM. Those used in Wi-Fi 6/7 are BPSK, QPSK, 16-QAM, 64-QAM, 256-QAM, 1024-QAM, and 4096-QAM. That used in Bluetooth 5/6 is GFSK. These are the methods, therefore, that we are going to briefly review in this chapter. For all but GFSK, the modulation is such that the modulated carrier is linearly related to the baseband signal, i.e., the data carrying signal at or near zero frequency. Such modulation is referred to as *linear modulation*. For GFSK, the modulated carrier is non-linearly related to the baseband signal thus the modulation is referred to as *non-linear modulation*. In this chapter, we first review the fundamentals of baseband transmission techniques, as these form the foundation upon which the techniques involving the modulation of an RF carrier are based. We then take a high-level view of the relevant linearly modulated methods followed by one of GFSK.

4.2 Baseband Data Transmission

Data transported by communication systems is typically a stream of pulses, random or very close to random in nature, and referred to as a *baseband* data stream. A stream of random, bipolar (a stream where the signal amplitude can be positive or negative), full-length rectangular pulses (a rectangular pulse of constant height for the entire pulse duration), where each pulse is equiprobable (equal probability of being positive or negative), is in fact a string of nonperiodic pulses. This is so because the value (polarity) of each pulse is entirely independent of all other pulses in the stream. For such a stream, of pulse duration τ, and hence symbol rate $f_s = 1/\tau$

© The Author(s), under exclusive license to Springer Nature Switzerland AG 2025

D. H. Morais, *5G/5G-Advanced, Wi-Fi 6/7, and Bluetooth 5/6,*
https://doi.org/10.1007/978-3-031-82830-0_4

symbols per second, the two-sided amplitude spectral density (amplitude of the signal as a function of frequency) is that of each of its nonperiodic pulses. The absolute value of its normalized positive (real) side is shown in Fig. 4.1. The form is (sin x)/x, where x is in radians and is well known, being referred to as the *sampling function* Sa(x). It occupies infinite bandwidth, with the first null at f_s. The spectrum within the first null is normally referred to as the main lobe. In communication systems, bandwidth is normally at a premium. Thus, the designer is motivated to filter the transmitted signal down to the minimum bandwidth possible without introducing errors into the transmission. Furthermore, at the receiver, the incoming signal is normally filtered to minimize the negative effects of noise and interference. Filtering the signal results in changes to the shape of the original pulses, spreading their energy into adjacent pulses. This spreading effect is known as *dispersion* and can result in distortion of the pulse amplitude at the sampling instant unless carefully controlled (the sampling instant is the instant at which the receiver decides on the polarity, and hence the binary value, of the received pulse). In the *Nyquist criterion* on bandwidth transmission, it is shown [1] that the minimum real channel bandwidth that independent symbols of rate f_s can be transmitted through, without resulting in symbol amplitude distortion at the sampling instant, is the Nyquist bandwidth $f_n = f_s/2$. Thus, for the rectangular pulse stream described above, the minimum transmission bandwidth is half the width of the main lobe. Because the stream is binary, one transmitted symbol contains one information bit. Thus, the bit rate f_b of the stream is the same as the symbol rate f_s and the minimum real transmission bandwidth f_n is $f_b/2$.

The essential components of a baseband digital transmission system are shown in Fig. 4.2. The input signal can be a binary or multilevel (>2) pulse stream. Such a system is referred to as one employing *Pulse Amplitude Modulation* (PAM). The transmitter low-pass filter, with transfer function $T(f)$, is used to limit the transmitted spectrum. Noise and other interference are picked up by the transmission medium and fed into the receiver filter. The receiver filter, with transfer function $R(f)$, minimizes the noise and interference relative to the desired signal. The output

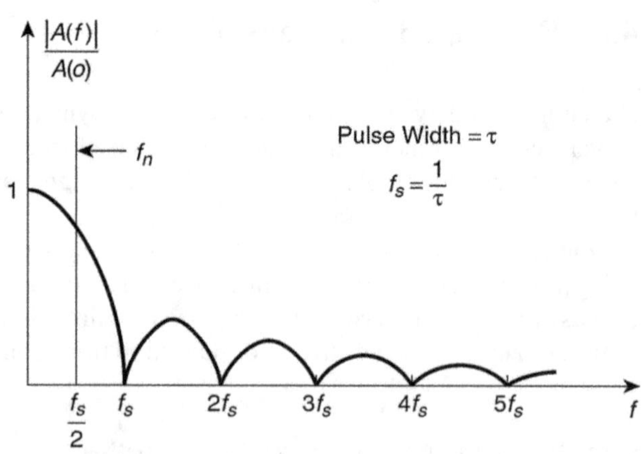

Fig. 4.1 The absolute normalized amplitude spectral density of a train of rectangular pulses

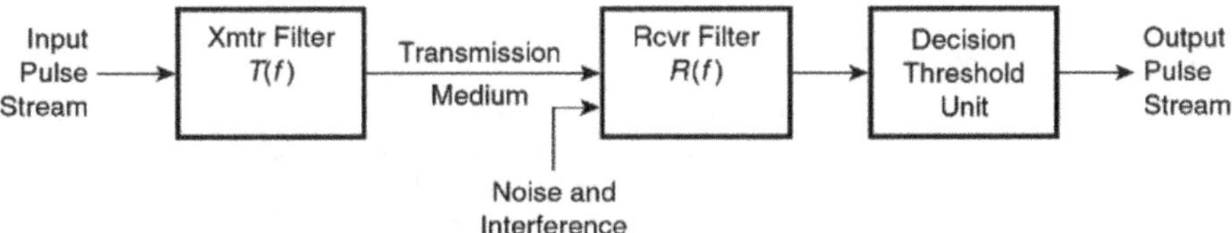

Fig. 4.2 Basic baseband digital transmission system

of the receiver filter is fed to a decision threshold unit which, for each pulse received, decides what was its most likely original level and outputs a pulse of this level. For a binary pulse stream, it outputs a pulse of amplitude $+V$ say, if the input pulse is equal to or above its decision threshold, which is 0 volts. If the input pulse is below 0 volts, it outputs a pulse of amplitude $-V$.

4.3   Linear Modulation

Above we discussed PAM baseband systems. Wireless communication systems, however, operate in assigned frequencies that are considerably higher than baseband frequencies. It is thus necessary to employ modulation techniques that shift the baseband data up to the operating frequency. In this section, we consider *linear modulation* systems.

These systems are the so-called because they exhibit a linear relationship between the baseband signal and the modulated RF carrier. We will commence this study by reviewing the so-called *double-sideband suppressed carrier (DSBSC)* modulation as this modulation forms the foundation on which many of the most widely used linear modulation methods are based.

4.3.1   Double-Sideband Suppressed Carrier (DSBSC) Modulation

A simplified DSBSC system for PAM signal transmission is shown in Fig. 4.3. First, a polar L-level PAM input signal, $a(t)$, with equiprobable symbols, is filtered with the low-pass filter, F_T, to limit its bandwidth to f_m say, and the filtered signal $b(t)$ applied to a multiplier. Also feeding the multiplier is a sinusoidal signal at the desired carrier frequency, f_c. As a result, the output signal of the multiplier, $c(t)$, is equal to $b(t) \times \cos 2\pi f_c t$.

$C(f)$, the amplitude spectral density of $c(t)$, as shown in Fig. 4.4, consists of two spectra. One is real, centered at f_c and the other imaginary, centered at $-f_c$, and each has a bandwidth $2f_m$ and an amplitude half that of $B(f)$, the amplitude spectral density of $b(t)$. As these spectra are symmetrically disposed on either side of the carrier

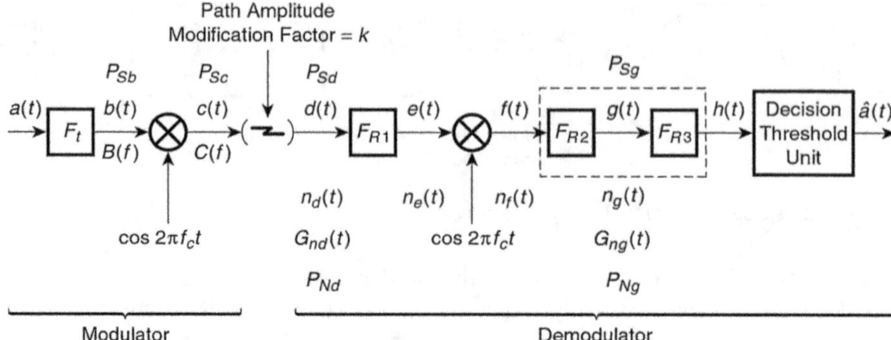

Fig. 4.3 Simplified one-way DSBSC system for PAM transmission

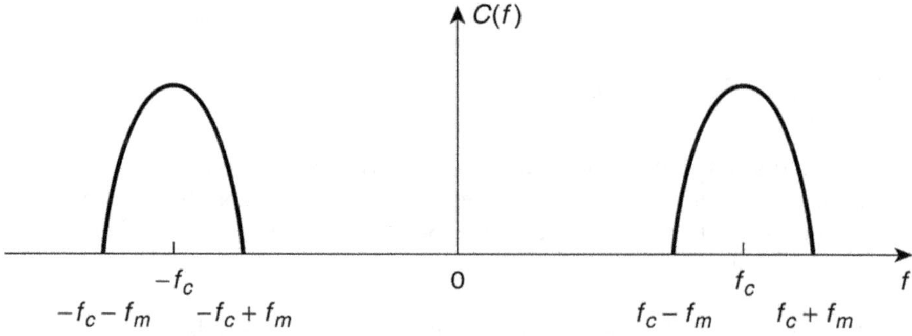

Fig. 4.4 DSBSC signal amplitude spectral density representation

frequency, the signal is referred to as a *double-sideband (DSB)* signal. Furthermore, as $b(t)$ is polar, and thus has no fixed non-zero average component, $c(t)$ contains no discrete carrier frequency component and is thus referred to as a *suppressed carrier* signal.

We assume that $c(t)$ travels over a linear transmission path and arrives at the demodulator input modified in amplitude by the factor k. Thus, the input signal $d(t)$ to the receiver is equal to $k \times c(t)$. This signal is passed through the bandpass filter F_{R1} to limit noise and interference. The bandwidth W of F_{R1} is normally greater than $2f_m$, the bandwidth of $d(t)$, so as to not impact the spectral density of $d(t)$. Assuming this to be the case, the output signal $e(t)$ of F_{R1} is equal to its input signal $d(t)$.

The signal $e(t)$ is fed to a multiplier. Also feeding the multiplier is the sinusoidal signal $\cos 2\pi f_c t$. Simple analysis shows that the output of the multiplier, $f(t)$, is given by

$$f(t) = kb(t)\cos^2\left(2\pi f_c t\right) \tag{4.1}$$

Substituting the trigonometric identity $\cos^2 x = \dfrac{1}{2}(1 + \cos 2x)$ into Eq. (4.1) we get

$$f(t) = \frac{k}{2}b(t) + \frac{k}{2}b(t)\cos(2 \cdot 2\pi f_c t)$$ (4.2)

Thus, by multiplying $e(t)$ by cos $2\pi f_c t$, a process referred to as *coherent detection*, we recover $b(t)$! We also create a second signal with the same double-sided bandwidth as $b(t)$ but centered at $2f_c$. The signal $f(t)$ is fed into the low-pass filter F_{R2} that eliminates the component of the signal centered about $2f_c$ while leaving the baseband component undisturbed. Thus, the output of F_{R2}, $g(t)$, is given by

$$g(t) = \frac{k}{2}b(t)$$ (4.3)

The signal $g(t)$ is fed to F_{R3} for final pulse shaping prior to level detection in the decision threshold unit. In practice F_{R2} and F_{R3} are combined into one but are shown separately here to add clarity to the analysis. The output, $\hat{a}(t)$, of the decision threshold unit is a PAM signal that is the demodulator's best estimate of the modulator input signal, $a(t)$.

The key point to note in the mathematics above is that it is the double multiplication of the original filtered signal $b(t)$ by $\cos 2\pi f_c t$ that leads to the recovery of $b(t)$ in the receiver. Furthermore, we note that double multiplication by $\sin 2\pi f_c t$ leads to the same detection result as $\sin^2 x = \frac{1}{2}(1 - \cos 2x)$.

4.3.2   *Binary Phase Shift Keying (BPSK)*

A special case of PAM transmission via a DSBSC system is when the PAM signal $a(t)$ in Fig. 4.3 has a binary, polar format. In this situation, if the filtered signal $b(t)$ has maximum peak amplitude of $\pm b$ volts say, then the modulated signal $c(t)$ varies between $c_1(t)$ and $c_0(t)$ as $b(t)$ varies between +b and –b, where

$$c_1(t) = b\cos 2\pi f_c t$$ (4.4)

$$c_0(t) = -b\cos 2\pi f_c t$$

$$= b\cos(2\pi f_c t + \pi)$$ (4.5)

When $b(t)$ is positive, the phase of $c(t)$ relative to the carrier phase is $0°$. When $b(t)$ is negative, the phase of $c(t)$ relative to the carrier phase is π radians or $-180°$. Thus, the relative phase has only two states. This modulation is referred to as *Binary Phase Shift Keying (BPSK)* and represents the simplest linear modulation scheme. Figure 4.5 shows typical examples of signals $a(t)$, $b(t)$, and $c(t)$. Figure 4.6 shows the *signal space* or *vector* or *constellation diagram* of $c(t)$. This diagram portrays

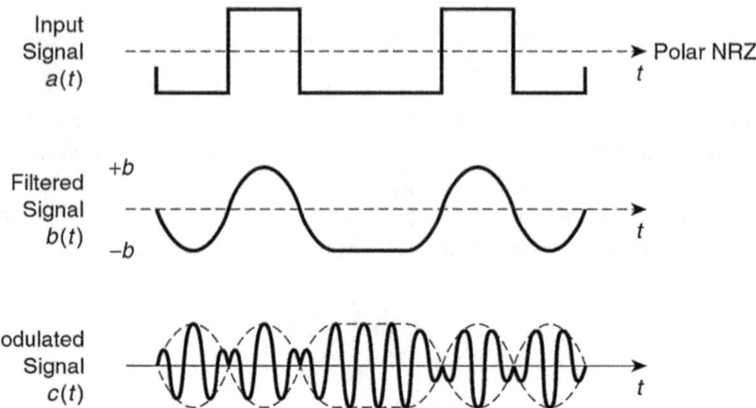

Fig. 4.5 Typical BPSK signals

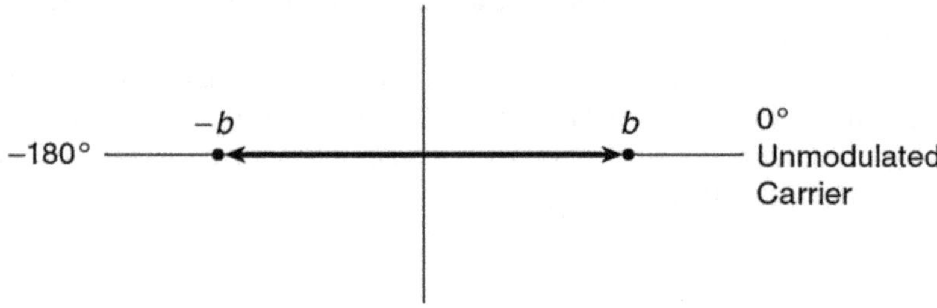

Fig. 4.6 Signal space diagram of BPSK modulated signal

both the amplitude and phase of $c(t)$ at the instances when the modulating signal $b(t)$ is at its peak.

Figure 4.7 shows the power spectral density (power of the signal as a function of frequency) of BPSK when the modulating signal is unfiltered. This power spectral density has the same $[\sin x/x]^2$ form as that of the two-sided baseband signal, except that it is shifted in frequency by f_c.

Also shown in Fig. 4.6 is the single-sided, double-sideband Nyquist bandwidth of the system, which is equal to f_b. Thus, at its theoretical best, BPSK is capable of transmitting 1 bit per second in each Hertz of transmission bandwidth. The system is therefore said to have a maximum *spectral efficiency* of 1 bit/s/Hz. Because filtering to achieve the Nyquist bandwidth is not practical, real BPSK systems have spectral efficiencies less than 1 bit/s/Hz. For a bandwidth in excess of the Nyquist by 25% say, then data at a rate of 1 bit/s requires 1.25 Hz of bandwidth, leading to spectral efficiency of 0.8 bits/s/Hz.

As we shall see in succeeding sections, spectral efficiencies much greater than that afforded by BPSK are easily realizable. As a result, when high data throughput is required, BPSK is rarely used in wireless communication networks, where, as a rule, available spectrum is limited and thus highly valued. Nonetheless, an

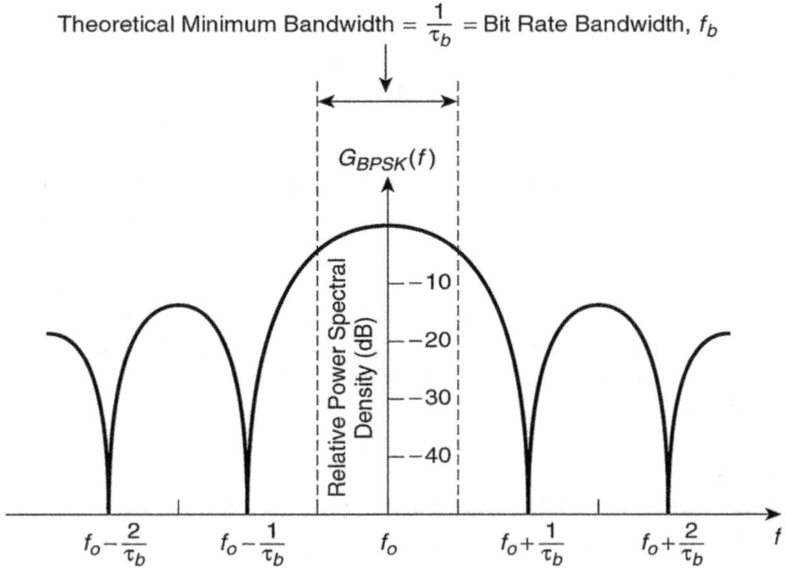

Fig. 4.7 Power spectral density of BPSK

understanding of its operating principles is very valuable in analyzing *Quadrature Phase Shift Keying* (QPSK), a popular modulation technique.

It can be shown [2] that the probability of bit error $P_{be(BPSK)}$ of a BPSK system, with optimum filtering, and in the presence of white Gaussian noise, is given by

$$P_{be(BPSK)} = Q\left[\left(\frac{2P_S}{P_N}\right)^{\frac{1}{2}}\right] \tag{4.6}$$

where P_S = the average signal power at the demodulator input, P_N = the noise power in the two-sided Nyquist bandwidth at the demodulator input, and Q is a well-tabulated probability function.

In Eq. (4.6) above, probability of error is defined in terms of a signal-to-noise ratio. In digital communication systems, however, it is just as common to define the probability of bit error, P_{be}, in terms of the ratio of the energy per bit E_b in the received signal to the noise power density N_0, i.e., the noise power in a 1-Hz bandwidth, at the receiver input. Defining P_{be} in terms of E_b/N_0 makes it easy to compare the error performance of different modulation systems for the same bit rate without taking bandwidth into account. For BPSK P_{be} in terms of E_b/N_0 is given by

$$P_{be(BPSK)} = Q\left[\left(2\frac{E_b}{N_0}\right)^{\frac{1}{2}}\right] \tag{4.7}$$

4.3.3 Pi/2 BPSK

Pi/2 BPSK is simply BPSK with a $\pi/2$ counter-clockwise phase shift rotation of the carrier frequency on every successive symbol. On the constellation diagram, BPSK occupies two phase positions, 0 and π radians, whereas Pi/2 BPSK occupies four-phase positions, namely, o, $\pi/2$, π, and $3/2\,\pi$ radians. To visualize phase transitions with Pi/2 BPSK, take, for example, the case where the current modulating symbol is of binary value 1 and the phase position is 0 radians. For the next symbol, there is an automatic $\pi/2$ phase rotation. Thus, if that next symbol were to also be of binary value 1, the next phase location would be $\pi/2$, but were it to be of binary value 0, the next phase location would be $3/2\,\pi$. Pi/2 BPSK exhibits the same bit error rate performance and the same spectral density characteristics as BPSK.

Pi/2 BPSK is sometimes used instead of BPSK because when the modulating data stream is filtered, as it usually is, it has a lower *peak-to-average power ratio* (PAPR) than BPSK. PAPR is important because linear systems require linear amplification and real transmitter output power amplifiers have a limit above which they cannot amplify linearly. The lower the PAPR the higher the average output power afforded.

4.3.4 Quadrature Amplitude Modulation (QAM)

The BPSK system described above is only capable of amplitude modulation accompanied by 0° or 180° phase shifts. However, by adding a *quadrature branch* as shown if Fig. 4.8, it becomes possible to generate signals with any desired amplitude and phase. In the quadrature branch, a second PAM baseband signal is multiplied with a sinusoidal carrier of frequency f_c, identical to that of the *in-phase* carrier

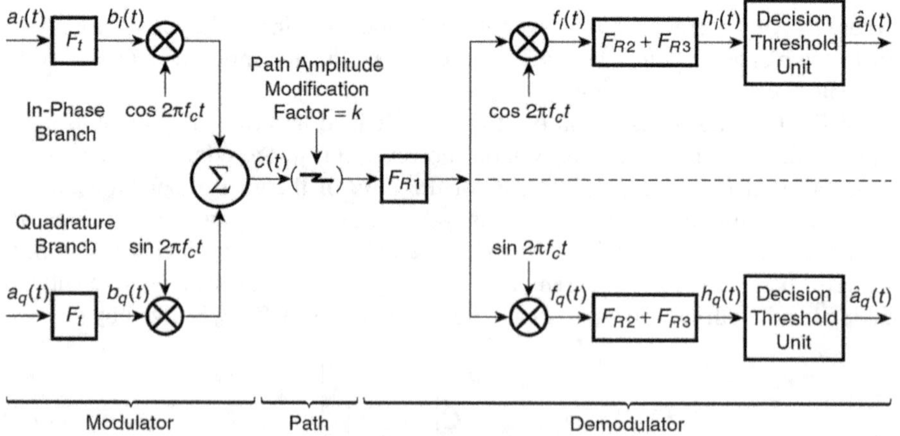

Fig. 4.8 Simplified one-way quadrature amplitude modulated system

but delayed in phase by 90°. The outputs of the two multipliers are then added together to form a *Quadrature Amplitude Modulated* (QAM) signal. At the QAM demodulator, the incoming signal is passed through the bandpass filter F_{R1} to limit noise and interference. It is then divided into two and each branch inputted to a multiplier, one multiplier being fed also with an in-phase carrier, $\cos 2\pi f_c t$, and the other with a quadrature carrier, $\sin 2\pi f_c t$.

Labeling the in-phase PAM filtered signal $b_i(t)$ and the quadrature PAM filtered signal $b_q(t)$, it can be shown that the output signal $f_i(t)$ of the in-phase multiplier is given by

$$f_i(t) = \frac{k}{2} b_i(t) + \frac{k}{2} b_i(t)\cos\left(2 \cdot 2\pi f_c t\right) + \frac{k}{2} b_q(t)\sin\left(2 \cdot 2 f_c t\right) \qquad (4.8)$$

The only difference between $f_i(t)$ and $f(t)$ of Eq. (4.2) for an in-phase only modulated system is the final component of Eq. (4.8). However, this component, like the second component in Eq. (4.8), is spectrally centered at $2f_c$ and is filtered prior to decision threshold detection, leaving only the original quadrature modulating signal $b_i(t)$.

It can also be shown that the output signal $f_q(t)$ of the quadrature multiplier, is given by

$$f_q(t) = \frac{k}{2} b_q(t) - \frac{k}{2} b_q(t)\cos\left(2 \cdot 2\pi f_c t\right) + \frac{k}{2} b_i(t)\sin\left(2 \cdot 2 f_c t\right) \qquad (4.9)$$

As with the output $f_i(t)$ from the in-phase multiplier, $f_q(t)$ consists of the original in-phase modulating signal $b_q(t)$ as well as two components centered spectrally at $2f_c$ which are filtered prior to decision threshold unit. Thus, by quadrature modulation, it is possible to transmit two independent bit streams on the same carrier with no interference of one signal with the other, given ideal conditions.

4.3.5   Quadrature Phase Shift Keying (QPSK)

Quadrature (or Quaternary) Phase Shift Keying (QPSK) is one of the simplest implementations of Quadrature Amplitude Modulation and is sometimes referred to as *4-QAM*. In it, the modulated signal has four distinct states. A block diagram of a conventional, simplified, QPSK system is shown in Fig. 4.9a. The binary non-return-to-zero (NRZ) input data stream $a_{in}(t)$, of bit rate f_b and bit duration τ_b, is fed to the modulator where it is converted by a serial to parallel converter into two NRZ streams, an I stream labeled $a_i(t)$ and a Q stream labeled $a_q(t)$, each of symbol rate f_B, half that of f_b, and symbol duration τ_B, twice that of τ_b. The relationship between the data streams $a_{in}(t)$, $a_i(t)$, and $b_q(t)$ is shown in Fig. 4.9b. The I and Q streams undergo standard QAM processing as described in Sect. 4.3.4. The in-phase multiplier is fed by the carrier signal $\cos 2\pi f_c t$. The quadrature multiplier is fed by the

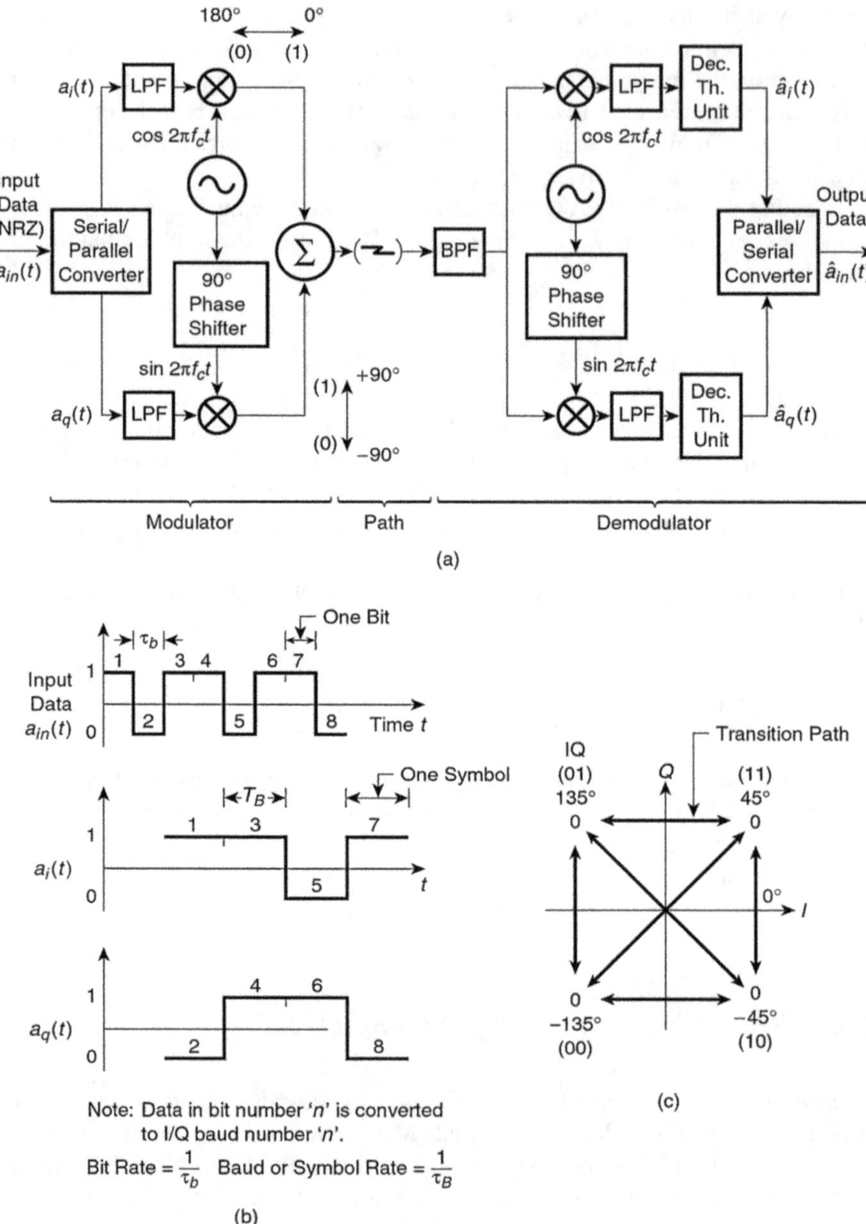

Fig. 4.9 QPSK system representation. (**a**) Block diagram. (**b**) Modular data streams. (**c**) Signal space diagram

carrier signal delayed by 90° to create the signal $\sin 2\pi f_c t$. The output of each multiplier is a BPSK signal. The BPSK output signal of the in-phase carrier-driven multiplier has phase values of 0° and 180° relative to the in-phase carrier, and the

BPSK output signal of the quadrature carrier-driven multiplier has phase values of 90° and 270° relative to the in-phase carrier. The multiplier outputs are summed to give a four-phase signal. Thus, QPSK can be regarded as two *associated BPSK* (ABPSK) systems operating in quadrature.

The four possible output signal states of the modulator, their *IQ* digit combinations, and their possible transitions from one state to another are shown in Fig. 4.9c. We note that either 90° or 180° phase transitions are possible. As an example, a 90° phase transition occurs when the *IQ* combination changes from 00 to 10, and a 180° phase transition occurs when the *IQ* combination changes from 00 to 11. For a system where $a_i(t)$ and $a_q(t)$ are unfiltered prior to application to the multipliers, phase transitions occur instantaneously, and thus the signal has a constant amplitude. However, for systems where these signals are filtered to limit the radiated spectrum, phase transitions occur over time, and the modulated signal has an amplitude envelope that varies with time. In particular, a 180° phase change results in a change over time in amplitude envelope value from maximum to zero and back to maximum.

In the demodulator, as a result of quadrature demodulation, signals $\widehat{a_i}(t)$ and $\widehat{a_q}(t)$, estimates of the original modulating signals are produced. These signals are then recombined in a parallel to serial converter to form $\widehat{a_{in}}(t)$, an estimate of the original input signal to the modulator.

As indicated above, QPSK can be regarded as two associated BPSK systems operating in quadrature. From a spectral point of view at the modulator output, two BPSK signal spectra are superimposed on each other. A graph of $G_{QPSK}(f)$ is shown in Fig. 4.10. We note that the widths of the main lobe and side lobes are half that for

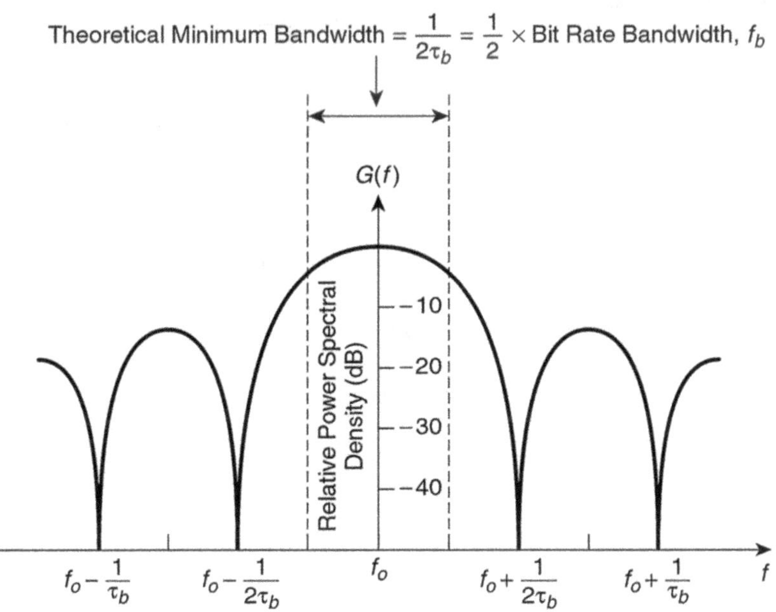

Fig. 4.10 Power spectral density of QPSK

BPSK given the same bit rate for each system. As a result, the maximum spectral efficiency of QPSK is twice that of BPSK, i.e., 2 bits/s/Hz.

It can be shown [2] that the probability of bit error $P_{be(QPSK)}$ of a QPSK system, with optimum filtering, and in the presence of white Gaussian noise, is given by

$$P_{be(QPSK)} = Q\left[\left(\frac{2E_b}{N_0}\right)^{\frac{1}{2}}\right] \tag{4.10}$$

We note that this relationship is identical to that for the probability of bit error versus E_b/N_0 for BPSK. Graphs of P_{be} [Bit Error Rate (BER)] versus E_b/N_0 for QPSK and other linear modulation methods are shown in Fig. 4.11.

In summary, for the same bit rate, the spectral efficiency of QPSK is twice that of BPSK with no loss in the probability of bit error performance in ideal circumstances. The QPSK hardware is, however, more complex than that required for BPSK. Furthermore, in transmission through non-linear components such as power amplifiers, filtered QPSK is subject to quadrature crosstalk, a situation where modulation on one quadrature channel ends up on the other. This situation can also arise in the receiver if the phase difference between the coherent detection oscillators is not kept to $90°$. Thus, in real-world environments, BPSK is a more robust modulation scheme than QPSK.

Fig. 4.11 BER versus E_b/N_0 for linear modulation methods

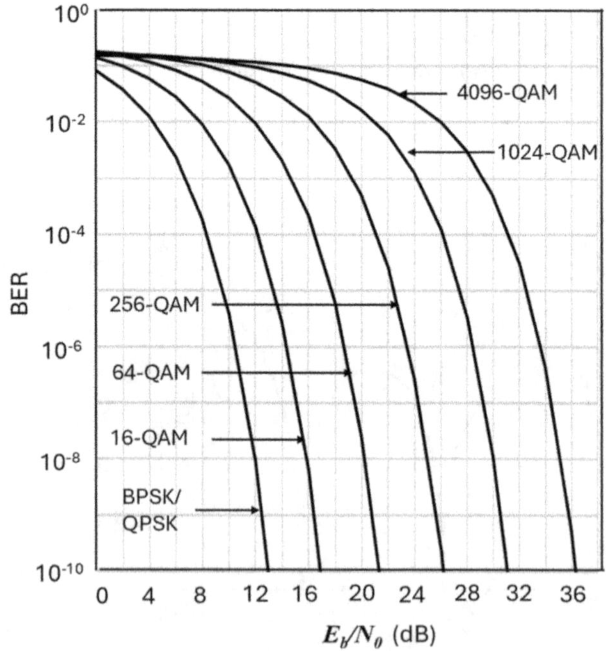

4.3.6   High-Order 2^{2n}-QAM

Though relatively easy to implement and robust in performance, linear four-phase systems such as QPSK do not often afford the desired spectral efficiency in commercial wireless systems. Higher-order QAM systems, however, do permit higher spectral efficiencies and have become very popular. A common class of QAM systems allowing high spectral density is one where the number of states is 2^{2n}, where n equals 2, 3, 4, …. A generalized and simplified block diagram of a 2^{2n}-QAM system is shown in Fig. 4.12. The difference between this generalized system and the QPSK system shown in Fig. 4.9 is that (a) in the generalized modulator, the I and Q signals $a_i(t)$ and $a_q(t)$ are each fed to a 2 to 2^n level converter prior to filtering and multiplication with the carrier and (b) in the generalized demodulator, the outputs of the decision threshold units are each fed to a 2^n to 2 level converter prior to being combined in a parallel to serial converter. 2^{2n}-QAM systems have been deployed commercially for values of n from 2 to 7.

For n equal 2, a 16-QAM system is derived. In such a system, incoming symbols to each modulator level converter are paired, and output symbols, in the form of signals at one of four possible amplitude levels, are generated in accordance with the coding table shown in Fig. 4.13a. The duration of these output symbols, τ_{B4L} say, is twice that of, τ_B, the duration of the input symbols. As a result of the application of the 4-level signals to the multipliers, the output of each multiplier is a 4-level amplitude modulated DSBSC signal, and the combined signal at the modulator output is a QAM signal with 16 states. Thus, 16-QAM can be treated as two 4-level PAM DSBSC systems operating in quadrature. The constellation diagram of a 16-QAM signal is shown in Fig. 4.13b. From this figure, it is clear that 16-QAM has an amplitude envelope that varies considerably over time, irrespective of whether the signal has been filtered or not and thus must be transmitted over a highly linear system if it is to preserve its spectral properties.

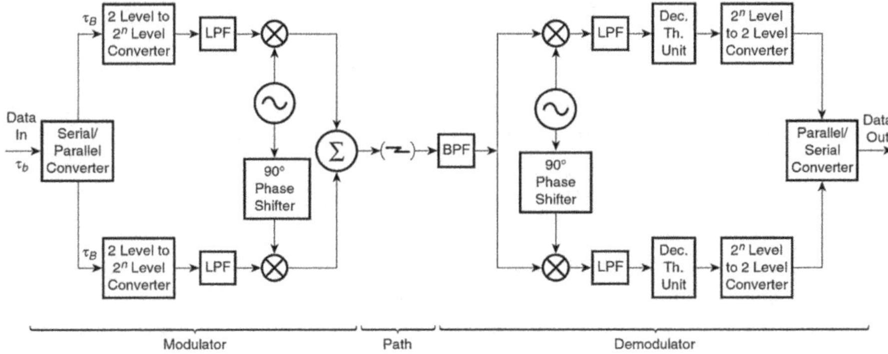

Fig. 4.12   Generalized block diagram of 2^{2n}-QAM system

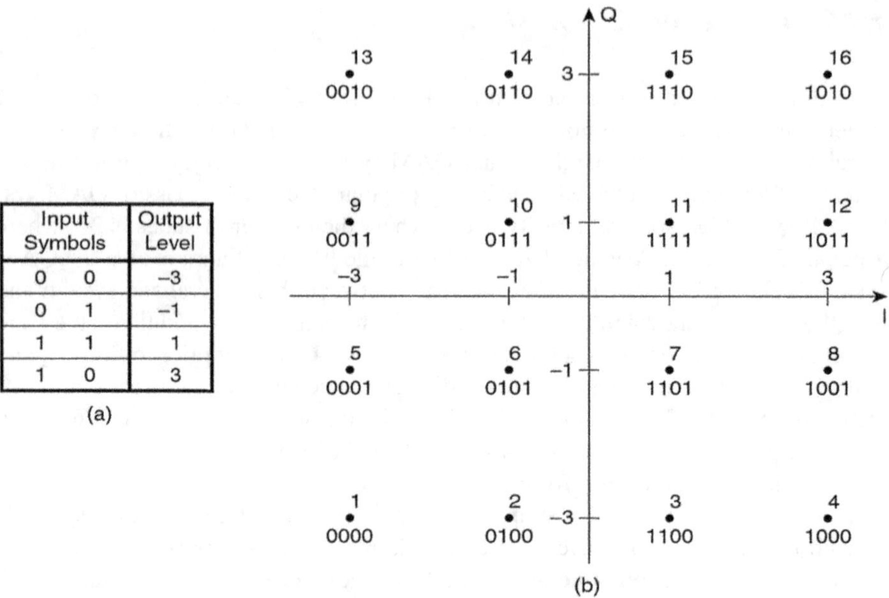

Fig. 4.13 16-QAM level converter coding table and constellation diagram. (**a**) 2-level to 4-level coding table. (**b**) Constellation diagram

$G_{16-QAM}(f)$ is such that its main lobes and side lobes are one-fourth as wide as those of BPSK. As a result, the maximum spectral efficiency off 16-QAM is 4 bits/s/Hz, twice that of QPSK.

For a 16-QAM system, it can be shown [2] that the probability of bit error $P_{be(16-QAM)}$ is given by

$$P_{be(16-QAM)} = \frac{3}{4} Q\left[\left(\frac{4}{5}\frac{E_b}{N_0}\right)^{\frac{1}{2}}\right] \tag{4.11}$$

A graph of $P_{be(16-QAM)}$ versus E_b/N_0 is shown in Fig. 4.11. It will be observed that for a probability of bit error of 10^{-3}, the E_b/N_0 required for 16-QAM is 3.8 dB greater than that required for QPSK. Thus, the doubling of the spectral efficiency achieved by 16-QAM relative to QPSK comes at the expense of probability of bit error performance.

The two-level to four-level coding shown in Fig. 4.13a is an example of *Gray coding*. In Gray coding, the bits that create any pair of adjacent levels differ by only one bit. It is interesting to note that, as a result of Gray coding in the *I* and *Q* channels, the 16 states shown in the signal space diagram in Fig. 4.13b are also Gray coded. Thus, an error resulting from one of these states being decoded as one of its closest adjacent states will result in only one bit being in error.

Using the same logic as applied to the above analysis of 16-QAM, but with generalized equations, it can be shown that the maximum spectral efficiency of 2^{2n}-QAM is $2n$ bits/s/Hz. Thus, that of 64-QAM is 6 bits/s/Hz, that of 256-QAM is 8 bits/s/Hz, and that of 1024-QAM is 10 bits/s/Hz.

It is shown in [2] that for Gray-coded 2^{2n}-QAM, with optimum filtering, and in the presence of white Gaussian noise, the generalized equation for probability of bit error P_e versus E_b/N_0 is

$$P_{be\left(2^{2n}-QAM\right)} = \frac{2}{\log_2 L}\left[1-\frac{1}{L}\right]Q\left[\left(\frac{6\log_2 L}{L^2-1}\frac{E_b}{N_0}\right)^{\frac{1}{2}}\right] \tag{4.12}$$

where $L = 2^n$.

The probability of error relationships for 64, 256, 1024-QAM and 4096-QAM are also shown in Fig. 4.11. We observe that as the number of QAM states increase, the P_e performance for the same bit rate decreases. An intuitive explanation of this is that, for the same average power, the area, on average, around each symbol in the constellation plane for correct symbol detection decreases as the modulation order increases. It thus takes less and less accompanying noise to result in a given probability that the location of the symbol plus noise falls outside the area of correct symbol detection. This translates into an increase in signal-to-noise ratio and hence an increase in the value of E_b/N_0 to maintain that given probability.

Table 4.1 presents spectral efficiency and E_b/N_0 for BER of 10^{-3} as a function of modulation method. Using the data shown therein, one can compare relative receiver signal power required for different modulation methods and different bit rates for a BER of 10^{-3}, a commonly used BER threshold level. Consider systems operating in a 23-MHz channel. Such a channel can allow a QPSK system with proper filtering to operate at a maximum bit rate of about 20 Mbits/sec. Here, the E_b/N_0 for an error rate of 10^{-3} is 6.8. If instead of using QPSK we change to 1024-QAM but keep the bit rate constant, then the E_b/N_0 required increases to 24.2, an increase of 17.4 dB. Since power is E_b x bit rate and we are keeping the bit rate constant, this implies that the received power must increase by 17.4 dB. This can only be achieved by increasing the transmitted power, lowering path loss by typically reducing path

Table 4.1 Spectral efficiency and E_b/N_0 for BER of 10^{-3} as a function of Modulation Method

Modulation method	Spectral efficiency Bits/s/Hz	E_b/N_0 for BER = 10^{-3}
BPSK	1	6.8
QPSK	2	6.8
16-QAM	4	10.5
64-QAM	6	14.8
256-QAM	8	19.5
1024-QAM	10	24.2
4096-QAM	12	29.2

distance, or a combination of both. Yes, the spectral efficiency is increased from 2 bits/s/Hz to 10 bits/s/Hz, i.e., a factor of 5, but this gives us no benefit as we have a 23-MHz channel available. One would thus not implement such a system in practice. What one may do, however, if one wanted a much higher data rate, is increase the data rate by a factor of 5 to 100 Mb/s by using 1024-QAM. Note, however, that this comes with an additional demand on received power as power is (E_b × bit rate) and the bit rate is now larger by a factor of 5, i.e., 7 dB. Thus, the power now required at the receiver is (17.4 + 7) = 24.4 dB larger than that required for a QPSK system operating at 20 Mbits/s.!! No free lunch.

In the QAM realizations discussed above, the I and Q carriers were modulated via first a parallel to series converter, followed by, for the cases where n was greater than1, a two-level to 2^n level converter. We note, however, that though helpful in conveying the modulation conceptually, such a physical realization is not necessary. All that is necessary is to utilize any mapping structure that converts each grouping of 2^n incoming data bits to the desired I and Q modulating values. Thus, with 16-QAM, for example, the mapper needs only be programmed to take any incoming 4-bit combination and map it to the I and Q values shown in Fig. 4.13b. For example, incoming bits 1110 are mapped to an I value of 1 and a Q value of 3. Viewed another way, it is mapped to the complex value $1 + j3$.

4.4   Non-linear Modulation: GFSK

In *frequency modulation* (FM), there is a non-linear relationship between the baseband modulating signal and the modulated carrier signal. Hence, all forms of FM modulation are non-linear. This non-linear relationship is such that individual frequency components of the modulating signal result in an infinite number of modulated signal sideband components. Thus, unlike the linear systems described above where the modulated carrier spectrum is simply the two-sided version of the modulating signal spectrum, here the spectrum has a complicated relationship relative to that at baseband.

Frequency Shift Keying (FSK) is a form of digital modulation whereby digital information is conveyed via the discrete changes in frequency to the carrier signal. The simplest form of FSK is binary FSK (BFSK) wherein a pair of discrete frequencies symmetrical about the carrier are used to transmit binary (0s and 1s) data. *Gaussian Frequency Shift Keying* (GFSK) is a modification of binary FSK (BFSK) and is the modulation method employed in Bluetooth communication.

With BFSK, the modulating data stream consists of nonperiodic rectangular pulses (see Fig. 4.5a) resulting in instantaneous changes to the carrier frequency which in turn widens the modulated signal spectrum. With GFSK, the rectangular stream of data pulses is filtered (see Fig. 4.5b) with a Gaussian filter prior to modulating the carrier. This filtering makes the frequency transitions smoother, resulting in a much narrower carrier spectrum and reduced interference to adjacent channels. This is achieved, however, at the expense of increased interference between demodulated data symbols. A Gaussian filter is used because such filters have the property of having no overshoot to a step function input while minimizing the rise and fall time. As a result, it provides the best combination of suppression of high frequencies while minimizing spatial spread. Figure 4.14 shows the amplitude frequency response of a Gaussian filter whose 3-dB bandwidth is 1 MHz.

For FSK systems, the *modulation index*, *m*, is given by

$$m = \Delta f \, / \, f_m \tag{4.13}$$

where Δf is the peak carrier frequency deviation and f_m is the maximum modulation frequency. In practice the value of f_m used is not normally the maximum modulating frequency but rather a frequency below which a large majority of the signal energy is contained. In GFSK systems, f_m is often taken to be B, the 3-dB bandwidth of the Gaussian filter. The specifications for GFSK as applied in Bluetooth are discussed in Chap. 10.

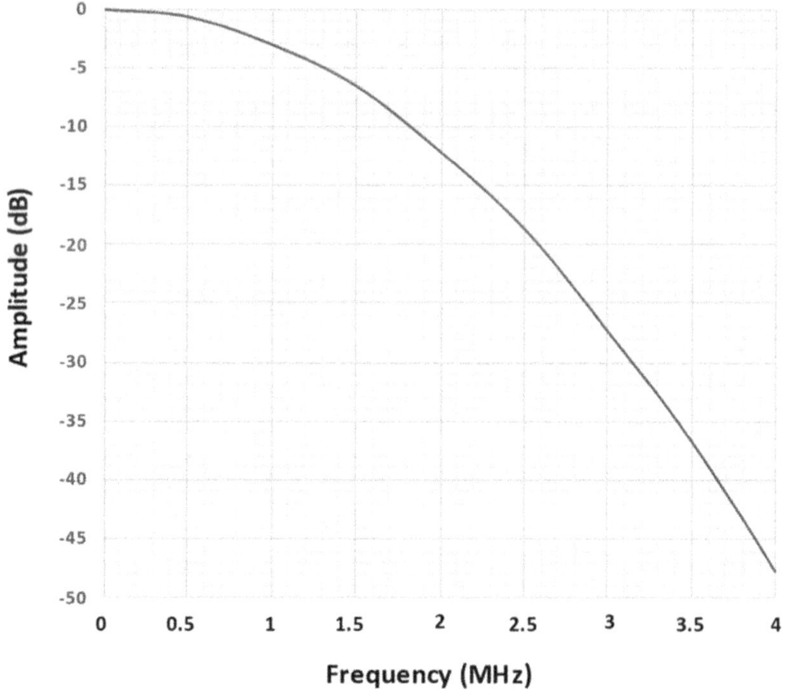

Fig. 4.14 Gaussian filter frequency response

4.5   Summary

The fundamental modulation methods used in 5G/5G-Advanced are Pi/2 BPSK, QPSK, 16-QAM, 64-QAM, 256-QAM, and 1024-QAM. Those used in Wi-Fi 6/7 are BPSK, QPSK, 16-QAM, 64-QAM, 1024-QAM, and 4096-QAM. That used in Bluetooth 5/6 is GFSK. These are the methods, therefore, that were briefly review in this chapter. For all but GFSK, the modulation is such that the modulated carrier is linearly related to the baseband signal, i.e., the data carrying signal at or near zero frequency. Such modulation is referred to as linear modulation. For GFSK, the modulated carrier is non-linearly related to the baseband signal thus the modulation is referred to as non-linear modulation. In this chapter, we first reviewed the fundamentals of baseband transmission techniques, then took a high-level view of the relevant linearly modulated methods followed by one of GFSK.

References

1. Feher K (1981) Digital communications: microwave applications. Prentice-Hall, Upper Saddle River
2. Morais DH (2004) Fixed broadband wireless communications: principles and practical applications. Prentice-Hall, Upper Saddle River

Chapter 5
Channel Coding and Link Adaptation

5.1 Introduction

Coding, in the binary communications world, is the process of adding a bit or bits to useful data bits in such a fashion as to facilitate the detection or correction or errors incurred by such useful bits as a result of their transmission over a non-ideal channel. Such a channel, for example, may be one that adds noise, or interference, or unwanted nonlinearities. In this chapter, we focus on error detection and error correction coding as applied to user data in 5G, 5G-Advanced, Wi-Fi 6/7, and Bluetooth 5/6, including *low-density parity check* (LDPC) coding, *binary convolution codes* (BCCs) coding, and *hybrid automatic repeat request* (HARQ). Error detection on its own obviously does nothing to improve error rate performance. However, as we shall see, it can aid in error correction when combined with other techniques.

Where helpful in understanding the various techniques presented, some basic coding theory and practice is presented. Several coding-related mathematical computations addressed below that involve binary data apply *modulo-2* arithmetic. Thus, following is a review of *modulo-2 addition, subtraction*, and *multiplication:*

Addition	Subtraction	Multiplication
$0 + 0 = 0$	$0 - 0 = 0$	$0 \times 0 = 0$
$0 + 1 = 1$	$0 - 1 = 1$	$0 \times 1 = 0$
$1 + 0 = 1$	$1 - 0 = 1$	$1 \times 0 = 0$
$1 + 1 = 0$	$1 - 1 = 0$	$1 \times 1 = 1$

D. H. Morais, *5G/5G-Advanced, Wi-Fi 6/7, and Bluetooth 5/6*,
https://doi.org/10.1007/978-3-031-82830-0_5

5.2   Error Detection: Cyclic Redundancy Check (CRC) Codes

Cyclic redundancy check (CRC) [1] is one of the most popular methods of error detection, being normally used with automatic repeat request (ARQ) (see Sect. 5.5 below) to facilitate error correction. It is employed by all three of the communication systems under study. To create a CRC codeword, a mathematical calculation is carried out on a block of useful binary data which is referred to as a *Frame*. In the calculation, a string of n 0s is appended to the Frame. This number is then divided by another binary number derived from the so-called *generator polynomial*. The number n is one less than the number of bits in the divisor. The remainder of this division represents content of the Frame and is added to the Frame. These added bits are called the *checksum*. At the receiver, the entire received codeword, i.e., the Frame plus the checksum, is divided by the same polynomial derived bits used by the transmitter. If the result of this division is zero, then the received Frame is assumed to be error free, if not, an error or errors is assumed to have occurred.

5.3   Forward Error Correction Codes

5.3.1   Introduction

Error control coding is a means of permitting the robust transmission of useful data by the deliberate introduction of redundancies into this data creating a codeword. One method of accomplishing this is to have a system that looks for errors at the receive end and, once an error is detected, makes a request to the transmitter for a repeat transmission of the codeword. In this method, called ARQ and alluded to above, a return path is necessary. Error correction coding that is not reliant on a return path inherently adds less delay to transmission. Such coding is referred to as *forward error correction* (FEC) coding. For digitally modulated signals, detected in the presence of noise, use of FEC results in the reduction of the residual BER, usually by several orders of magnitude and a reduction of the receiver 10^{-6} threshold level by about one to several dBs depending on the specific scheme employed. Figure 5.1 shows typical error performance characteristics of an uncoded versus FEC coded digitally modulated system. The advantage provided by a coded system can be quantified by *coding gain*. The coding gain provided by a particular scheme is defined as the reduction in E_b/N_0 in the coded system compared to the same system but uncoded for a given BER and the same data rate. Coding gain varies significantly with BER, as can be seen from Fig. 5.1, and above a very high level may even be negative. As BER decreases, the coding gain increases until it approaches a limit

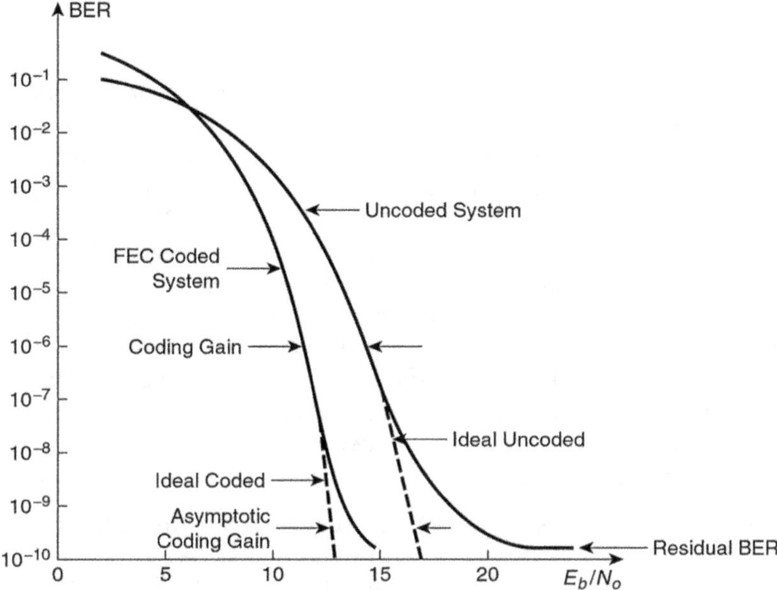

Fig. 5.1 Typical error performance of an uncoded versus FEC coded system

as the BER approaches zero (zero errors). This upper limit is referred to as the *asymptotic coding gain* of the coding scheme.

FEC works by adding extra bits to the bit stream prior to modulation according to specific algorithms. These extra bits contribute no new message information. However, they allow the decoder, which follows the receiver demodulator, to detect and correct, to a finite extent, errors as a result of the transmission process. Thus, improvement in BER performance is at the expense of an increase in transmission bit rate. The simplest error detection-only method used with digital binary messages is the parity check scheme. In the even-parity version of this scheme, the message to be transmitted is bundled into blocks of equal bits, and an extra bit is added to each block so that the total number of 1s in each block is even. Thus, whenever the number of 1s in a received block is odd, the receiver knows that a transmission error has occurred. Note, however, that this scheme can detect only an odd number of errors in each block. For error detection and correction, the addition of several redundant (check) bits is required. The number of redundant bits is a function of the number of bits in errors that are required to be corrected.

FEC codes can be classified into two main categories, namely, *block codes* and *convolution codes*. A LDPC code is a block code. A BCC is a convolution code. For the transmission of user data, 5G NR used LDPC coding. Wi-Fi 6 uses LDPC or BCC coding, and Bluetooth 5 uses, in some instances, BCC coding. We will first study block coding in general followed specifically by LDPC coding then move on to BCC coding.

5.3.2   Block Codes

In *systematic binary linear* block encoding, the input bit stream to be encoded is segregated into sequential message blocks, each k bits in length. The encoder adds r check bits to each message block, creating a codeword of length n bits, where $n = k + r$. The codeword created is called an (n, k) block codeword, having a block length of n and a coding rate of k/n. Such a code can transmit 2^k distinct codewords, where, for each codeword, there is a specific mapping between the k message bits and the r check bits. The code is *systematic* because, for all codewords, a part of the sequence in the codeword (usually the first part) coincides with the k message bits. As a result, it is possible to make a clear distinction in the codeword between the message bits and the check bits. The code is *binary* because its codewords are constructed from bits and *linear* because each codeword can be created by a linear modulo 2-addition of two or more other codewords. The following simple example [2] will help explain the basic principles involved in linear binary block codes.

Example 5.1 The Basic Features and Functioning of a Simple Linear Binary Code
Consider a (5, 2) block code where three check bits are added to a 2-bit message. There are thus four possible messages and hence four possible 5-bit encoded codewords. Table 5.1 presents the specific choice of check bits associated with the message bits. A quick check will confirm that this code is linear. For example, codeword 1 can be created by the modulo 2-addition of codewords 2, 3, and 4. How does the decoder work? Suppose codeword 3 (10011) is transmitted, but an error occurs in the second bit so that the word 11,011 is received. The decoder will recognize that the received word is not one of the four permitted codewords and thus contains an error. This being so, it compares this word with each of the permitted codewords in turn. It differs in four places from codeword 1, three places from codeword 2, one place from codeword 3, and two places from codeword 4. The decoder therefore concludes that it is codeword 3 that was transmitted, as the word received differs from it by the least number of bits. Thus, the decoder can detect and correct an error.

The number of places in which two words differ is referred to as the *Hamming distance*. Thus, the logic of the decoder in Example 5.1 is, for each received word, select the codeword closest to it in Hamming distance.

Block decoding can be accomplished with *hard decision decoding*, where the demodulator outputs either ones or zeros as in Example 5.1. Here, the codeword

Table 5.1 A (5,2) block code

Codeword #	Codeword	
	Message bits	Check bits
1	00	000
2	01	110
3	10	011
4	11	101

chosen is the one with the least Hamming distance from the received sequence. However, decoding can be improved by employing *soft decision decoding*. With such decoding the demodulator output is normally still digitized, but to greater than two levels, typically eight or more. Thus, the output is still "hard" but more closely related to the analog version and thus contains more information about the original sequence. Such decoding can be accomplished in a number of ways. One such way is to choose as the transmitted codeword the one with the least *Euclidian distance* from the received sequence. The Euclidian distance between sequences is, in effect, the root mean square error between them.

5.3.3   Classical Parity Check Block Codes

Before we describe the features of LDPC codes, we review some of the features of classical parity check block codes. In such a code, each codeword is of a given length, n say, contains a given number of information bits, k say, and a given number of *parity check bits*, r say, and thus $r = n—k$. The structure can be represented by a *parity check matrix* (PCM), where there are n columns representing the digits in the codeword, and r rows representing the equations that define the code. Consider one such code, where the length n is 6, the number of information bits k is 3, and hence the number of parity bits r is 3. The rate of this code is thus $k/n = 3/6 = 1/2$. We label the information bits V1 to V3 and the parity bits V4 to V6. The parity check equations for this code are shown in Eq. (5.1) below, where + represents modulo 2-addition:

$$V1 + V2 + V4 = 0$$
$$V2 + V3 + V5 = 0 \tag{5.1}$$
$$V1 + V2 + V3 + V6 = 0$$

To determine the relationship between information bits and parity check bits, the constraints of Eq. (5.1) can be rewritten as:

$$V4 = V1 + V2$$
$$V5 = V2 + V3 \tag{5.2}$$
$$V6 = V1 + V2 + V3$$

Using Eq. (5.2), then, for example, the message bits 110 produces the parity check bits

$$V4 = 1 + 1 = 0$$
$$V5 = 1 + 0 = 1$$
$$V6 = 1 + 1 + 1 = 0$$

and thus, the codeword for this message is 110,010.

The equations of (5.1) can be represented in matrix form, as shown in Fig. 5.2a, where each equation maps to a row of the matrix. This matrix is referred to as the PCM associated with Eq. (5.1).

Equation 5.1 can also be represented in the graphical form. When done, such a graph is referred to as a *Tanner graph*. A Tanner graph is a bipartite graph, i.e., a graph which contains nodes of two different types and lines (also referred to as edges) which connect nodes of different types. The bits in the codeword form one set of nodes, referred to as *variable nodes* (VNs), and the parity check equations that the bits must satisfy form the other set of nodes, referred to as the *check nodes* (CNs). The Tanner graph corresponding to the PCM matrix above is shown in Fig. 5.2b.

Errors can be detected within limits in any received codeword by simply checking if it satisfies all associated parity check equations. However, block codes can only detect a set of errors if errors don't change one codeword into another.

To not only detect bit errors but to also correct them, the decoder must determine which codeword was most likely sent. One way to do this, as mentioned above, is to choose the codeword closest in minimum distance to the received codeword. This method of decoding is called *maximum likelihood* (ML) *decoding*. For codes with a short number of information bits, this approach is feasible as the computation required is somewhat limited. However, for codes with thousands of information bits in a codeword such as can be found in 5G NR and Wi-Fi 6, the computation required becomes too excessive and expensive. For such codes, alternative decoding methods have been devised and will be discussed below.

5.3.4   *Low-Density Parity Check (LDPC) Codes*

Low-density parity check (LDPC) codes are employed in 5G NR and Wi-Fi 6 systems for the transmission of user data. They are linear FEC codes and were first proposed by Gallager [3] in his 1962 Ph.D. thesis. LDPC codes can provide higher coding gains and lower error floors than binary convolution codes (BCCs) and a

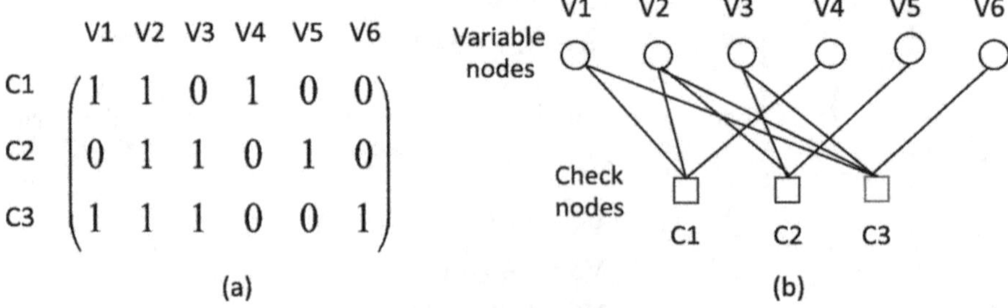

Fig. 5.2 A PCM and associated Tanner graph. (**a**) PCM. (**b**) Tanner graph

more complex version of such codes called convolution turbo codes. LDPC codes are distinguished from other parity check codes by having parity check matrices where the percentage of 1s is low, i.e., of low density, hence the nomenclature. This sparseness of 1s results in a decoding complexity which increases only linearly with code length.

LDPC codes are said to be either regular or irregular. A LDPC code is regular if all the VNs have the same degree, i.e., they are connected to the same number of CNs, and all the CNs have the same degree, i.e., they are connected to the same number of VNs. When this is the case, then every code bit is contained in the same number of equations, and each equation contains the same number of code bits. Irregular codes relax these conditions, allowing VNs and CNs of different degrees. Irregular codes, it has been found, can provide better performance than regular ones.

Figure 5.3a shows the PCM of a simple LDPC where $n = 12$, and Fig. 5.3b shows the associated Tanner graph. It will be noted that this is a regular LDPC code, where from the PCM perspective each code bit is contained in three equations and each equation involves four code bits and from the Tanner graph perspective, each bit node has three lines connecting it to parity nodes, and each parity node has four lines connecting it to bit nodes. We note that in the PCM, there are 108 positions in all of which only 36, or 33%, are ones.

5.3.4.1   Encoding of Quasi-Cyclic LDPC Codes

5G NR and Wi-Fi 6 use a class of LDPC codes called *Quasi-cyclic* (QC) LDPC codes. These codes are used in support of channels that transmit user data in both downlink and uplink directions. With these codes, encoding and decoding hardware implementation tends to be easier than with other type of LDPC codes, achieving this without measurably degrading the relative performance of the code. The PCM of a QC LDPC code is defined by a small graph, called a *base graph* or *protograph* and the key to high performance of a QC LDPC code is the construction of the base graph. The PCM consists of an array of block matrices, each such matrix

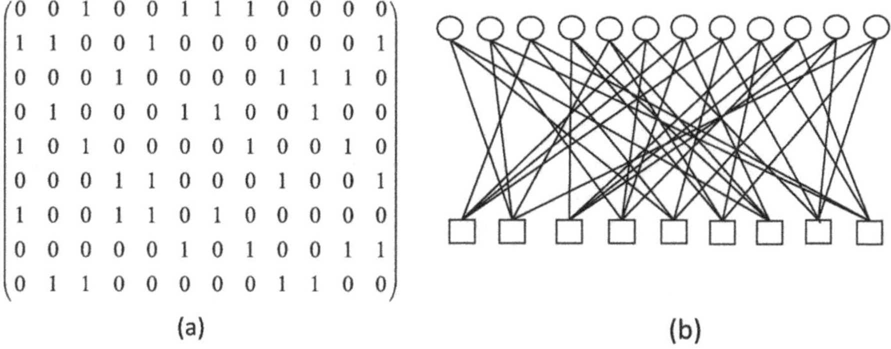

(a)   (b)

Fig. 5.3 PCM and Tanner graph for an LDPC code where $n = 12$. (**a**) PCM. (**b**) Tanner graph

corresponding to an element of the base graph, and the structure of each block matrix is determined by the value of the associated base graph element. The size of the individual block matrix determines the size of the PCM, thus by making the block matrix large we can create a large PCM. Below is an example of a very simple base graph U and an associated PCM H. The block matrices are created by cyclically shifting a so-called 3×3 identity matrix. For a $Z \times Z$ identity matrix, Z is referred to as the lifting factor as the larger it is, the larger the size of H.

$$U = \begin{pmatrix} 3 & -1 & -1 & 1 \\ -1 & 0 & -1 & 2 \end{pmatrix}, \qquad H = \begin{pmatrix} 1 & 0 & 0 & 0 & 0 & 0 & 0 & 0 & 0 & 0 & 1 & 0 \\ 0 & 1 & 0 & 0 & 0 & 0 & 0 & 0 & 0 & 0 & 0 & 1 \\ 0 & 0 & 1 & 0 & 0 & 0 & 0 & 0 & 0 & 1 & 0 & 0 \\ 0 & 0 & 0 & 1 & 0 & 0 & 0 & 0 & 0 & 0 & 0 & 1 \\ 0 & 0 & 0 & 0 & 1 & 0 & 0 & 0 & 0 & 1 & 0 & 0 \\ 0 & 0 & 0 & 0 & 0 & 1 & 0 & 0 & 0 & 0 & 1 & 0 \end{pmatrix}$$

5G NR LDPC codes specify two base graphs in order to cover a large range of information payloads and rates:

- *Base graph 1* (BG1) is specified to accommodate large payloads and high rates. It allows the handling a maximum information payload of 8448 bits and a nominal code rate is 8/9, but by adding additional parity bits to the base graph, the rate can be decreased to as low as 1/3 without puncturing or repetition. This low-rate aids in permitting extended coverage and the meeting of the cell-edge throughput goal of 100 MB/s.
- *Base graph 2* (BG2) is specified to accommodate smaller payloads and lower rates. It allows the handling a maximum information payload of 3840 bits and nominal code rates that vary between 2/3 and 1/5 without puncturing or repetition. The choice between base graph 1 and base graph 2 is a function of transport block size and the code rate chosen for the initial transmission.

The Wi-Fi 6 standard defines (QC) LDPC codes with codeword block lengths of 648, 1296, and 1944 with lifting factors of 27, 54, and 81, respectively. For each codeword, block length code rates of 1/2, 2/3, 3/4, 5/6 are defined.

5.3.4.2 Decoding of LDPC Codes

A big distinguishing feature between LDPC codes and classical block codes is how they are decoded. Unlike classical codes that are usually of short length and decoded via ML decoding, LDPC are decoded iteratively using *message-passing* algorithms [4] since their functioning can be described as the passing of messages along the lines of the Tanner graph. Each node on the Tanner graph works in isolation, having access only to the information conveyed by the lines connected to it. The message-passing algorithms create a process where the messages pass back and forth between the bit nodes and check nodes iteratively. For optimum decoding the messages passed are estimates of the probability that the codeword bit information passed is

1. Each estimate is in the form of a *Logarithmic Likelihood Ratio* (LLR), where, for a codeword bit b_i, LLR$(b_i) = \ln$ prob.$(b_i = 0) - \ln$ prob.$(b_i = 1)$. A positive LLR indicates a greater confidence that the associated bit is of value 0, while a negative LLR indicates a greater confidence that the bit value is 1. The magnitude of the LLR expresses the degree of confidence. Decoding as described above is termed *belief propagation* decoding and proceeds as follows:

1. Each codeword is outputted from the channel not as hard outputs (1s or 0s) but rather as soft outputs. These soft outputs are converted into initial estimates in the form of LLRs.
2. Each bit node sends its initial estimate to the check nodes on the lines connected to it.
3. Each check node makes new estimates of the bits involved in the parity equation associated with that node and sends these new estimates via the connecting lines back to the associated bit nodes.
4. New estimates at the bit nodes are sent to the check nodes and process steps 3 and 4 repeated until a permitted codeword is found or the maximum number of permitted iterations reached.

5.3.5   *Binary Convolution Coding*

In binary convolution coding, a binary message stream is encoded in a continuous fashion, rather than from message block to message block as in block coding. A generalized convolution encoder is shown in Fig. 5.4. It consists of a K stage shift register and n modulo-2 adders. Register outputs (though not necessarily all) feed adders, with the choice of which register output feeds which adders determining the specifics of the encoder output. At each shift instant, 1 bit is shifted into the first stage of the register and all bits already in the register are shifted 1 stage to the right. The outputs of the n adders are then sequentially sampled by the

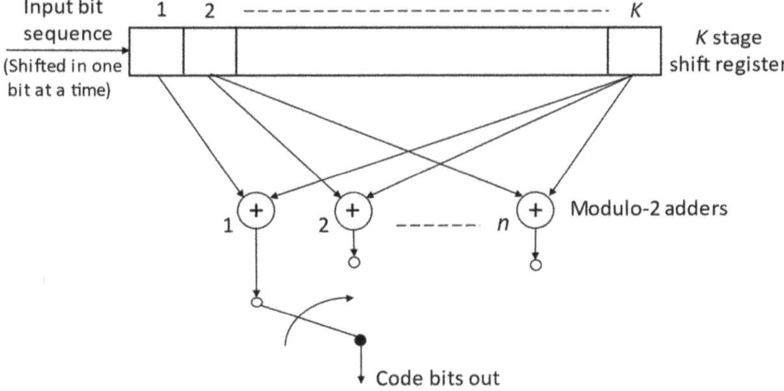

Fig. 5.4 Generalized convolution coder

commutator to generate *n code bits*. The sampling is repeated for each inputted bit. *K* is referred to as the *constraint length* of the encoder and signifies the number of shifts, 1 bit at a time, over which a single information bit can influence the output of the encoder. Since for each input of 1 message bit there are *n* code bits, the code rate is 1/*n*. The choice of connections between the shift registers and the adders required to yield good error correction properties is complicated and does not lend itself to straightforward solutions. However, computer searches have yielded good codes.

Not all encoders are structured as shown in Fig. 5.4. In some encoders, a connection or connections is/are made to the wire feeding the first shift register. Thus, the bit on the wire influences the output of the encoder. The result is that the constraint length in this situation is now $K + 1$.

Convolution encoders are said to possess states. The states represent the system memory. The convention used here is that the current bit is located in the leftmost stage, not on the wire feeding that stage. It should be noted, however, that is also common convention to show the current bit as being on the input feed. Thus, in reviewing an encoder, it is important to establish the convention being used. With the convention being applied here, when 1 new bit is shifted into the register, what's then in "memory" are the additional $(K - 1)$ bits in the register. The resulting *n* code bits emitted from the encoder are not only a function of the inputted bit, but also the $(K - 1)$ bits in memory. The number of states of the encoder equals the number of combinations of the bits in memory and thus equals $2^{(K-1)}$.

To help in the understanding the functioning of convolution encoders, consider the very simple encoder shown in Fig. 5.5. As there is a two-stage shift register, the constraint length *K* is equal to 2, the encoder possesses two states, 0 and 1, and as there are two output bits for each input bit the rate is 1/2. In this encoder only one adder is used. Labeling the input bits as the *i* bits, the output bits out of the second register as the *l* bits, and those out of the adder as the *m* bits, we have.

$$l_1 = i_{1-1} \tag{5.3}$$

$$m_j = i_j + i_{j-1} \left(modulo - 2\, addition \right) \tag{5.4}$$

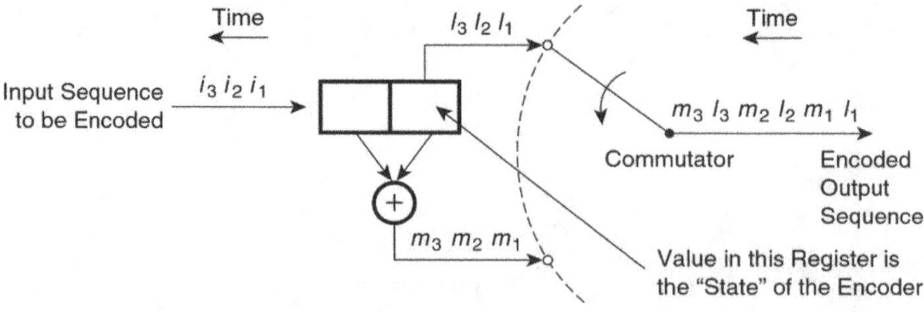

Fig. 5.5 Simple convolution coder

In Fig. 5.5, for the purpose of clarity, the i, l, and m bits are shown in a way to represent the order of their movement into and out of the encoder. We assume that the register at the start contains all zeros. To demonstrate the functioning of the encoder with a real input, consider the input sequence $i_1 = 1$, $i_2 = 0$, $i_3 = 1$, $i_4 = 1$. Table 5.2 presents the register contents before and after each bit in the sequence is inputted. It also shows the resulting output sequences l_j and m_j as dictated by applying Eqs. (5.3) and (5.4) to the "new register contents." Thus, for the input bit sequence 1011, we get at the output of the commutator, the output code bit sequence 01 11 01 10.

To determine the output sequence for a given input sequence, it is not always necessary to go through the process given in Table 5.2. Instead, the output sequence can be determined with the aid of a so-called *trellis diagram*. The trellis diagram for the encoder shown in Fig. 5.5 is shown in Fig. 5.6. In this diagram, if the input bit is a 0 we follow the solid line, if a 1, the dashed line. Above each line is the corresponding two outputs' bits resulting from the injection of the input bit. Note that each horizontal line corresponds to one of the two states of the encoder and that the trellis continues ad infinitum. Once an encoding trellis is fully formed, it repeats itself in each succeeding time interval.

Decoding of convolution codes is quite complex, being accomplished by either *sequential decoding* [5, 6] or *maximum likelihood sequence detection* (*MLSD*) [5, 6]. The latter, better known as *Viterbi decoding,* is the more elegant and the more easily accomplished and, hence, the more popular of the two. As there are no distinct codewords, the decoder decides between possible code sequences. In essence, the Viterbi algorithm, with a knowledge of the encoder trellis, finds the path through this trellis that most closely resembles the received signal sequence. Since the paths through the trellis represent all possible transmit sequences, the algorithm searches every path through the trellis. However, by recognizing that at a given transition time, only one path to each given node on the trellis may be the correct one, it discards the unlikely paths at every node, keeping only the path that is closest to the received sequence. It thus restricts the number of paths under consideration to a manageable level. The path that is kept is called the *survivor path*. The algorithm used to determine the survivor path is most easily understood by way of an example.

Table 5.2 Relationship between input sequence, register contents, and output sequence for encoder shown in Fig. 5.5

Input bit number	Input bit value	Old register contents	New register contents	Output sequence l_j m_j
1	1	00	10	0 1
2	0	10	01	1 1
3	1	01	10	0 1
4	1	10	11	1 0

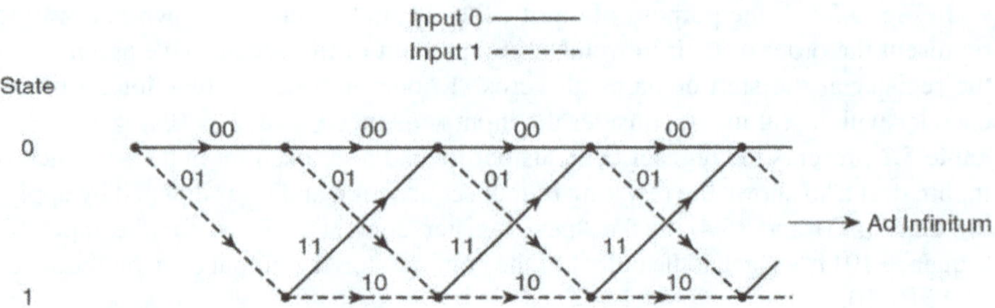

Fig. 5.6 Trellis diagram for encoder in Fig. 5.5

Table 5.3 Encoder input, errored demodulator output example

Time	t_1	t_2	t_3	t_4
Encoder input	1	0	1	1
Encoder output	01	11	01	10
Demodulator output/decoder input	01	01	01	10

Bit in error

Example 5.2 Demonstration of the algorithm to determine the survivor path
Let us assume that the input sequence 1011 that we considered earlier is coded via
the encoder of Fig. 5.5 and sent over a noisy channel and, as a result, there is one
error in the demodulated output, and hence decoder input, as given in Table 5.3. The
application of this decoding algorithm to the encoder trellis of Fig. 5.6, when the
decoder input is as given in Table 5.3, is shown in Fig. 5.7. The decoder works as
follows:

At t_2, it computes the distance between the received code bits 01 and labels of the
two branches leading to the two nodes shown at that time. The computed distance is
shown in Fig. 5.7a above each label, and the total distance from t_1 is shown at the
t_2 nodes.

At t_3 it computes the distance between the received code 01 and the labels of the four
branches leading to the two nodes at that time. It then adds these distances to the distances at the start of the branches. At each node, it discards that branch with the higher
total distance metric. The discarded branches are shown "crossed out" in Fig. 5.7b.

An attempt is made to repeat the same process at time t_4, but now we have a
problem. In Fig. 5.7c, we observe that on the upper t_4 node both total metrics are
identical, each being equal to two. When this happens, the rule is to arbitrarily throw
out one of the two branches. Applying this rule, we throw out the upper branch.
Note that in this figure we don't show previously discarded branches.

At t_5, we apply the process again. It results, as shown in Fig. 5.7d, in the upper of
the two.

branches to each node having the higher total metric distance and hence being
discarded. However, note what else happens,. Since the branches discarded both

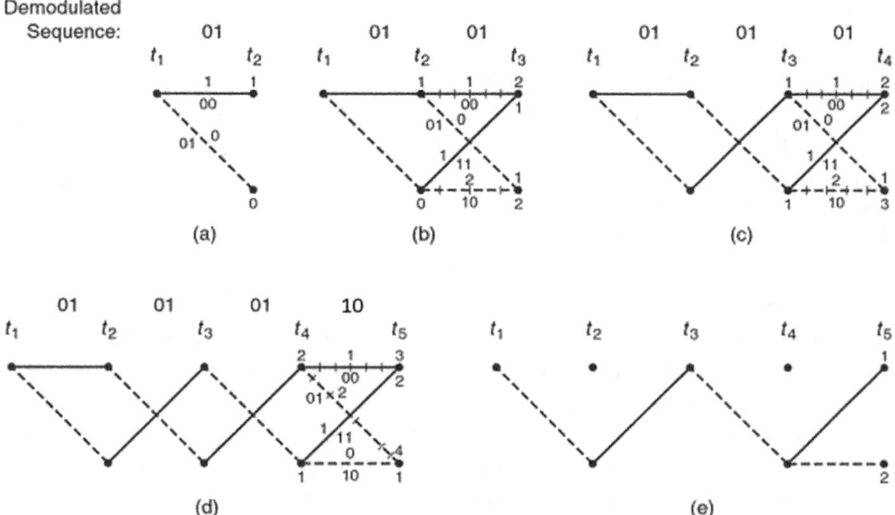

Fig. 5.7 Viterbi decoding example based on the trellis in Fig. 5.6 and demodulated output as per Table 5.3

originated at the upper $t4$ node, clearly the chain of branches that feeds only $t4$ can also be discarded. As a result, we end up with the surviving paths shown in Fig. 5.7e, where from t_1 to t_4 there is only one common path. However, the final decoding rule says that if the deletion process leaves only one survivor path over some earlier data period, the input bit that would have led to that path is outputted as the decoded result. Thus, for the period t_1 to t_2, the trellis is a dashed line and the output is therefore a 1. Similarly, for the period $t2$ to $t3$, we find that the output is 0 and for t_3 to t_4 it is 1. We cannot yet decode the fourth bit. More input bits are required to accomplish this.

In summary, we decoded correctly the first three bits encoded and sent, even though there was an error in one of the bits of the received sequence.

Example 5.2, though instructive, is based on a highly simplified encoder of constraint length K equal to 2. In practice encoders have constraint lengths of 3 or more, leading to more states and hence many more paths to process. In fact, the complexity of the Viterbi decoding increases exponentially with the code's constraint length. However, the larger the value of K, the larger the coding gains that are achievable. As a result, in most practical encoders, K is usually less than about ten in value. As with block codes, decoding of convolution codes can be improved via the use of soft decision decoding.

Convolution codes do not perform well in channels that result in error bursts. However, as we shall see in Sect. 5.6, this limitation can be overcome by applying a process called block interleaving.

BCC is employed in both Wi-Fi 6 and Bluetooth 5 communication systems.

5.3.6 Puncturing

For convolution coders of the form shown in Fig. 5.4 the bit rate is $1/n$. Thus, the highest rate achievable is ½. It is possible, however, to increase the coding rate of a rate ½ coder by employing a technique called *puncturing* and codes produced using this technique are called punctured codes. Puncturing is the process of discarding some of the bits of an error correction codeword prior to transmission. To understand how puncturing works, consider the case where we want to create a rate ¾ code from a rate ½ code. For the rate ½ code, for every three 3-bit input sequence, we have a 6-bit output. To create the rate ¾ code, we simply delete 2 of the 6 output bits, thus giving us 4 output bits for every 3 input ones. The performance of this punctured code is dependent on which bits were deleted. The rate ¾ punctured code can be decoded using same decoder as required for original unpunctured rate ½ code. To use the rate ½ decoder, the rate ¾ punctured code is transformed back into a rate ½ structure by simply inserting dummy symbols (1s or 0s) into positions where bits were deleted before decoding. The dummy bits result in an impairment of the rate ½ code correcting capability. However, the impaired capability is normally no less than that which would have been achieved had an unpunctured rate 3/4 code been employed in the first place (an unpunctured rate ¾ code would require feeding bits in and shifting bits three at a time and 3 K shift registers). Punctured codes allow the dynamic selection of code rates using just one basic encoding/decoding structure. Such dynamic selection may be appropriate in situation where the channel condition varies significantly, and a variable transmission rate is acceptable. Under a good channel condition, a high-rate option is selected but, if the channel condition deteriorates to the point where the BER approaches an unacceptable level, the coding rate is reduced, thus restoring the BER to an acceptable level.

Puncturing is employed in both 5G NR and Wi-Fi 6 systems.

5.4   Block Interleaving

Block interleaving is a technique applied to mitigate the impact of error bursts. The number of coded blocks (codewords) involved in the interleaving process is referred to as the *interleaving depth*. The larger the interleaving depth, the longer the burst of contiguous errors that can be corrected but the greater the delay introduced. A block interleaver consists of a structure that supports a two-dimensional array, of width equal to the codeword length and depth equal to the interleaving depth. Data is fed into the interleaver row by row until the array is full and then read out column by column, resulting in a permutation of the order of the data. At the receive end of data transmission, the original data sequence is restored by a corresponding de-interleaver. A common designation for an interleaver is π and that for its corresponding de-interleaver π^{-1}. A simple example will illustrate how *block interleaving* addresses error bursts in a simple block code.

Example 5.3 How block interleaving impacts the decoding of signals corrupted with error bursts

In this example, an encoder with an interleaving depth of five creates the original 4-bit codewords shown in Fig. 5.8a. These codewords are fed to an interleaver that creates the interleaved words shown in Fig. 5.8b. Assume these interleaved words are then transmitted over a wireless noisy channel. As a result, a burst of five contiguous errors appears on the demodulated interleaved words, as indicated in Fig. 5.8b. Note, however, that when deinterleaved as shown in Fig. 5.8c, the five errors are now spread over the five original codewords. Assuming a decoder that can correct just one bit per codeword, it can, nonetheless, decode the original five codewords without error. Without interleaving, a burst of five contiguous errors would have caused errors in two codewords, which would have been beyond the capability of the decoder to eliminate.

Block interleaving is employed in both 5G NR and Wi-Fi 6 systems.

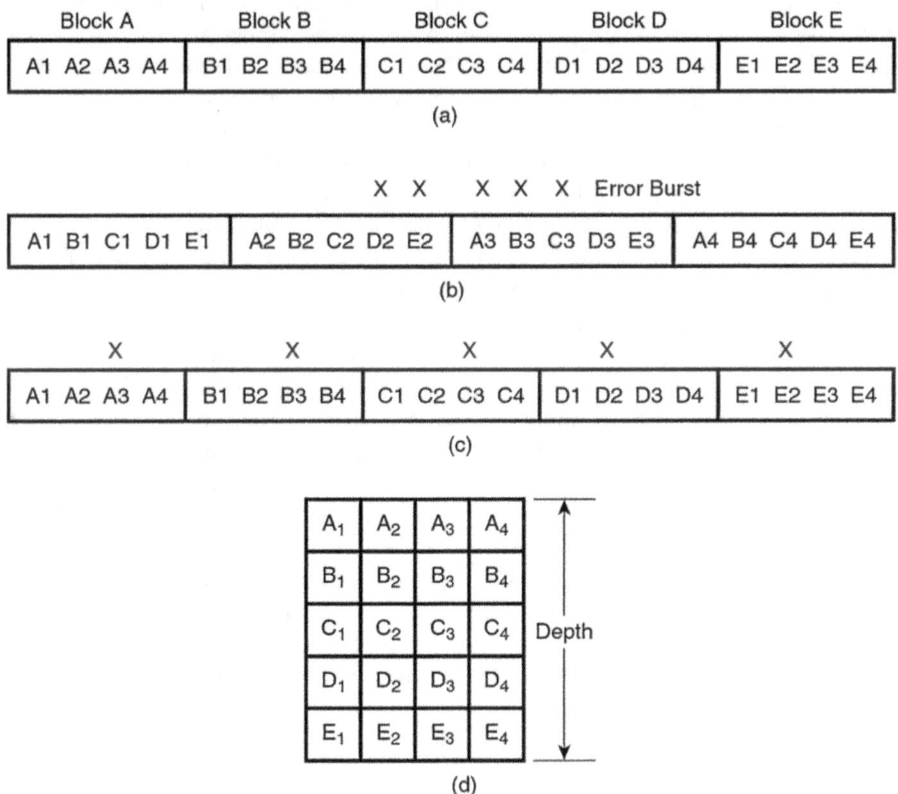

Fig. 5.8 Block interleaving. (**a**) Original codewords. (**b**) Interleaved words. (**c**) Deinterleaved words. (**d**) Two-dimensional array

5.5   Automatic Repeat Request (ARQ)

The *automatic repeat request* (ARQ) error control method is used with packet data systems where data transmission is not a continuous stream. In this method, the receiver detects packet errors, typically via CRC, but makes no attempt to correct packets received in error. Rather, if a packet error is not detected, an ACK message sent to transmitter, which then sends the next packet. However, if a packet error is detected, a NACK message is sent to the transmitter which then retransmits the packet in error. ARQ error control is not very efficient, since if just one bit in a codeword is in error, a retransmission of the entire codeword is required.

5.6   Hybrid ARQ (HARQ)

Hybrid ARQ (HARQ) is an error control scheme that employs both FEC and ARQ. By combining advantages of both these schemes, much better performance than ARQ is achieved, especially in time-varying fading channels and channels experiencing fluctuating interference. There are two main types of HARQ schemes: type-I and type-II. In the type-I scheme, the receiver discards any packet received in error that it is unable to correct and requests a retransmission. This process continues until an error-free packet is available or until a maximum number of retransmission attempts have taken place. In the type-II scheme, which is the more efficient of the two, any packet that cannot be successfully decoded is saved for possible further use, and a retransmission is requested. There are two forms of type-II scheme: *chase combining* and *incremental redundancy*.

Chase combining (CC) is the simpler form of the type-II scheme. Here, when an error is detected at the receiver, a NACK is sent back to transmitter, but the packet in error stored in a buffer memory is not discarded. If the retransmitted packet, which is an identical copy of original, is again in error, then it is combined with previous packet in an attempt to correct the error. New packets are transmitted with the same scheme being applied until either the error is corrected, or the maximum number of retries specified is reached.

Incremental redundancy (IR), the more complex form of the type-II scheme, is similar to CC, but here each retransmitted packet contains a different mix of information bits and parity bits. Transmitted packets are identified by version. The first packet, which typically contains all the information bits in addition to some parity bits, is referred to as redundancy version 0 or RV0 and is followed by other redundancy versions.

The combining schemes used in CC and IR where, as we have seen, the receiver combines the received signals from multiple transmission attempts, are called "soft-combining" schemes, and bits combined are called "soft bits." HARQ with incremental redundancy is the primary way of handling retransmissions in 5G NR.

5.7 Link Adaptation

In a mobile cell, the further a *mobile unit* (MU) is from the *base station* (BS) the lower the average signal level received by both the MU and the BS. In addition, the further the MU is from the BS, the larger the fading that is likely to occur on the BS/MU path. If a high-order modulation, for example, 256-QAM, is used on the transmission from one MU and a low-order from another, for example, QPSK, the MU with the high-order modulation will be able to send more data in a given bandwidth than the one with the low-order modulation. However, at the BS, the received power from the high-level MU must be greater than that from the low-level one for similar BER performance, as BER performance decreases as modulation level increases. This suggests that for similar BER performance on individual MU to BS links, modulation should be function of path length, assuming equal transmitter power on all MUs. High-level modulation could be employed on short paths, low-level on long ones, and thus the greater throughput the shorter the path. For scheme to be truly effective, however, it should be implemented in downstream direction as well.

In *adaptive modulation*, modulation is adjusted automatically per BS/MU link, independently on downstream and upstream directions, so as to optimize trade-off between capacity and reach. Modulation is typically adjusted on a burst-by-burst basis. In addition to optimizing throughput and coverage area, adaptive modulation also helps in combating co-channel interference. It does this by decreasing modulation complexity and hence increasing interference resistance whenever performance starts to degrade due to such interference. Instantaneous measurements are made of the SNR and carrier-to-interference ratio and modulation complexity varied dynamically. For example, if a MU has good SNR because its path is not fading, more complex modulation is used. If, on the other hand, the BS has a poor carrier-to-interference ratio because its received signal is fading, and an interferer is present, less complex modulation is used.

An improvement on adaptive modulation is *adaptive modulation and coding* (AMC), where not only is modulation adjusted dynamically but coding as well. It allows the current modulation-coding scheme to be matched to current channel conditions for each user. With AMC, the power of the transmitted signal is held constant over a frame interval, and the modulation-coding format changed burst-by-burst within the frame to match users current received signal conditions. To effect AMC, each transmitter must have knowledge of the path condition; hence feedback from receiver to transmitter is required.

In a mobile system, a difference is inevitable between channel status when reported (feedbacked) and when applied. HARQ is very effective in addressing the difference problem. Thus, AMC combined with HARQ provides robust transmission over a time-varying channel.

AMC is employed in both 5G NR and Wi-Fi 6 systems.

5.8 Summary

In this chapter, we focused on error detection and error correction coding as applied in 5G/5G-Advanced, Wi-Fi 6/7, and Bluetooth 5/6, including the low-density parity check (LDPC) codes, binary convolution codes (BCC), and hybrid automatic repeat request (Hybrid ARQ). Such detection and correction are crucial to achieving bit probability of error performance required of the systems under study. For more detailed information on channel coding see [7].

References

1. Peterson WW, Brown DT (1961) Cyclic codes for error detection. In: Proceedings of the IRE, vol 49, Jan 1961
2. Burr A (2001) Modulation and coding for wireless communications. Pearson Education, Harlow
3. Gallager RG (1963) Low density parity codes. MIT Press, Cambridge, MA
4. Mackay DJC (1999) Good error-correcting codes based on very sparse matrices. IEEE Trans Inf Theory 45:399–431
5. Sklar B (2001) Digital communications: fundamentals and applications. Prentice Hall PTR, Upper Saddle River
6. Burr A (2001) Modulation and coding for wireless communications. Pearson Education, Edinburgh Gate
7. Morais DH (2021) Key 5G physical layer technologies, 2nd edn. Springer, Cham

Chapter 6
Channel Usage Techniques

6.1 Introduction

In this chapter, we review the channel usage techniques employed by 5G/5G-Advanced, Wi-Fi 6/7, and Bluetooth 5/6. In the cases of 5G and Wi-Fi channel, usage is via multi-carrier-based multiple-access techniques where multi-carrier implies communication via a plurality of radio-frequency carriers and multiple accesses implies two-way communication between a central station and multiple surrounding stations. First, we review orthogonal frequency division multiplexing (OFDM). OFDM is a multi-carrier technique that lends itself to form the basis of multiple-access techniques. One such technique is employed by both 5G and Wi-Fi, namely orthogonal frequency division multiple access (OFDMA). Another such technique is employed optionally by 5G, namely discrete Fourier transform spread OFDM (DFTS-OFDM). Having reviewed OFDM, next addressed are OFDMA and DFTS-OFDM. In the case of Bluetooth, channel usage is via frequency hopping spread spectrum (FHSS). This technique is therefore introduced.

6.2 Orthogonal Frequency Division Multiplexing (OFDM)

6.2.1 OFDM Basics

Orthogonal frequency division multiplexing (OFDM) is not a modulation technique though it is often loosely referred to as such. Rather, it is a *multi-carrier transmission technique*, which allows the transmission of data on multiple adjacent subcarriers, each subcarrier being modulated in a traditional manner with a linear modulation scheme such as QAM. In OFDM, the data for transmission is, via a serial to parallel converter, converted into several parallel streams, and each stream

D. H. Morais, *5G/5G-Advanced, Wi-Fi 6/7, and Bluetooth 5/6*,
https://doi.org/10.1007/978-3-031-82830-0_6

used to modulate a separate subcarrier. Thus, only a small amount of the total data is transmitted via each subcarrier, in a subchannel a fraction of the width of the total channel. As a result, in a multipath fading environment, as a fade notch moves across the channel, the fading appears to each subchannel almost as a flat fade. Figure 6.1 demonstrates such a scenario. The fade thus induces a significantly reduced amount of *intersymbol interference* (ISI) compared to that which would be experienced by a single carrier modulated system with a spectrum extending across the entire band. Furthermore, while those subchannels at or close to the notch may experience a deep fade and hence thermal noise and ISI-induced burst errors, those removed from the notch will not. In the reconstructed original data stream, these burst errors are randomized, due to the interleaving which results from the parallel to serial process and therefore more easily corrected with FEC. The robustness of OFDM to multipath interference is one of its most important properties.

Figure 6.2 shows the spectrum of a standard four-channel *frequency division multiplex* (FDM) system. In such a scheme, modulated signals are stacked adjacent to each other with a guard band between each adjacent spectrum to ensure that there is no overlap between signals and to facilitate recovery via filtering of each signal at the receive end. While this approach works well, it suffers from the major drawback that, because of the spacing required between subcarriers, it wastes spectrum. As a result, it requires more bandwidth than would be required by a single carrier modulated by the original data stream, assuming that the same modulation is applied to the single carrier as to the subcarriers. With OFDM, however, the subcarriers are cleverly stacked close to each other. This results in overlapping spectra which (1) eliminates the spectral utilization drawback without incurring an adjacent intersubcarrier interference penalty and (2) retains advantages in the multipath arena that accrue to parallel transmission of lower data rate streams. In general, for the same basic modulation method and same data rate, OFDM leads to better bandwidth efficiency and hence higher data capacity compared to standard FDM as well as single carrier transmission. It achieves the latter because its spectrum below the main lobe of its first subcarrier and that above the main lobe of its last subcarrier falls off faster than that of a single carrier.

OFDM achieves its close stacking property, without adjacent channel interference, by making the individual subcarrier frequencies *orthogonal* to each other.

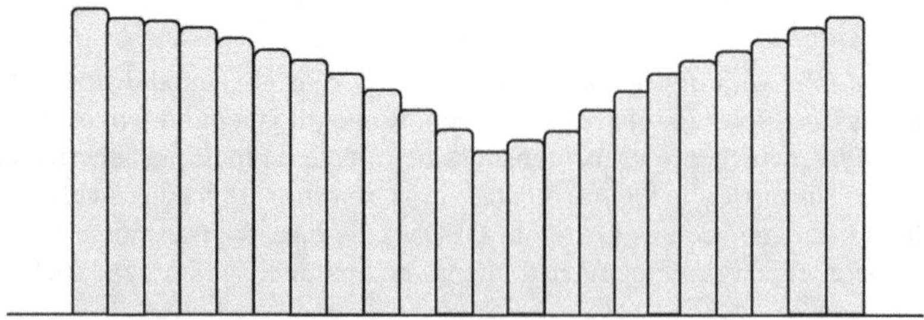

Fig. 6.1 Multipath fade across a multi-carrier signal

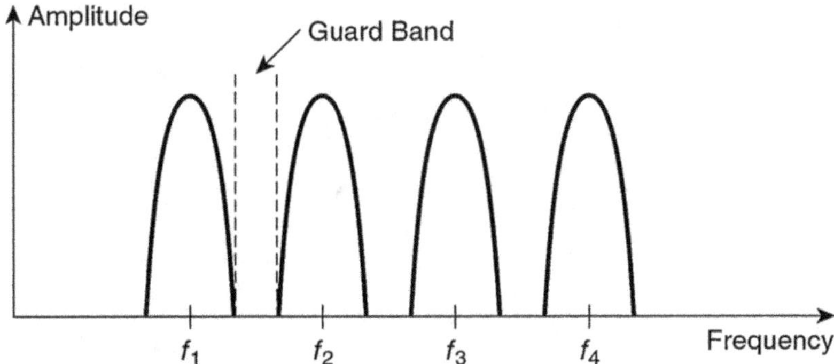

Fig. 6.2 Standard FDM frequency spectrum

This is accomplished by having each subcarrier frequency be an integer multiple of the symbol rate of the modulating symbols and each subcarrier separated from its nearest neighbor(s) by the symbol rate. Thus, the multiple to generate each carrier is one integer different from those to generate its adjacent neighbors. Figure 6.3a shows an example of four subcarriers in the time domain over one symbol period, τ. Figure 6.3b shows these subcarriers in the frequency domain when each is modulated with symbols of period τ in the form of rectangular pulses. With such modulation, each subcarrier spectrum has the familiar sin x/x format, but note that, by the choice of subcarrier frequencies, the spectra overlap and each spectrum has a null at the center frequency of each of the other spectra in the system. Why this structure allows the individual modulated subcarriers to be demodulated with no interference from its neighbors is not intuitively obvious, at least not to the author, by simply looking at Fig. 6.3b. Remember, we are looking at a frequency domain representation, not a time domain representation of overlapping pulses. If anything, this figure seems to represent the ultimate in adjacent channel interference. The explanation here, like the devil, is in the detail, the detail being orthogonality. The term orthogonal here refers to the total uncorrelation between variables and here is used in reference to the mathematical relationship between the subcarriers [1].

In order to avoid the construction of a large number of subchannel modulators in the transmitter and an equal number of demodulators in the receiver, modern OFDM systems utilize *digital signal processing (DSP)* devices. In fact, it is the availability of such devices that have made the commercialization of OFDM possible. Directly as a consequence of the orthogonality of the OFDM signal structure, modulation is able to be performed, in part, by using DSP to carry out an *inverse discrete Fourier transform* (IDFT) [1]. The IDFT transforms a signal in the form of discrete complex numbers from the frequency domain to the time domain. Similarly, demodulation is able to be performed, in part, by using DSP to carry out a *discrete Fourier transform* (DFT) [1] which transforms a signal in the form of discrete complex numbers from the time domain to the frequency domain. The conventional Fourier transform relates to continuous signals. However, digital signal processing is based on signal samples and so uses IDFT and DFT, which is a variant of the conventional

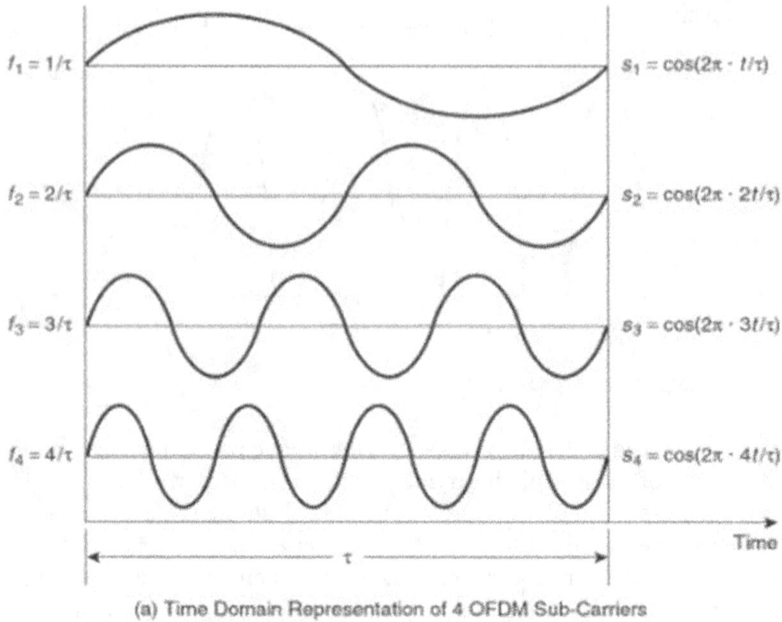

(a) Time Domain Representation of 4 OFDM Sub-Carriers

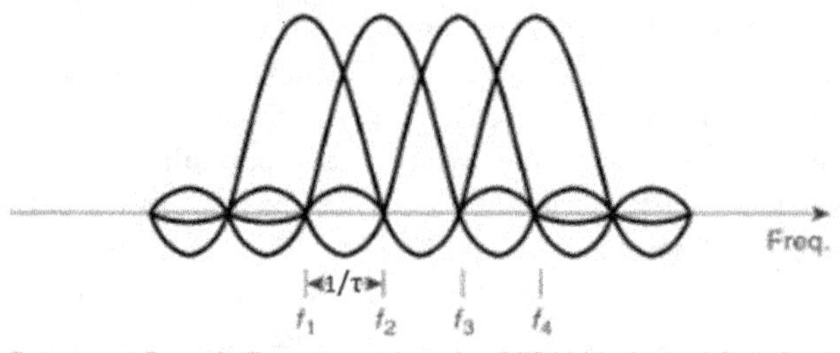

(b) Frequency Domain Representation of 4 OFDM Modulated Sub-Carriers

Fig. 6.3 Time and frequency representations of OFDMj subcarriers. (**a**) Time domain representation of four OFDM subcarriers. (**b**) Frequency domain representation of four OFDM modulated subcarriers

transform. In fact, it is typically the *inverse fast Fourier transform* (IFFT) and the *fast Fourier transform* (FFT) that are normally applied, these being a rapid mathematical method for computer applications of IDFT and DFT, respectively.

Figure 6.4 shows the basic processes in an IFFT/FFT-based OFDM system. The incoming serial data is first converted from serial to parallel in the S/P converter. If there are N subcarriers, N sets of parallel data streams are created. Each set contains a subset of parallel data streams, depending on the type of modulation. For example, if the modulation is 16-QAM, then each set contains four parallel data streams, the four bits in each symbol period of these streams being used to define a specific point in the 16-QAM constellation. The parallel data streams feed the mapper. For each subcarrier, the input data per symbol period is mapped

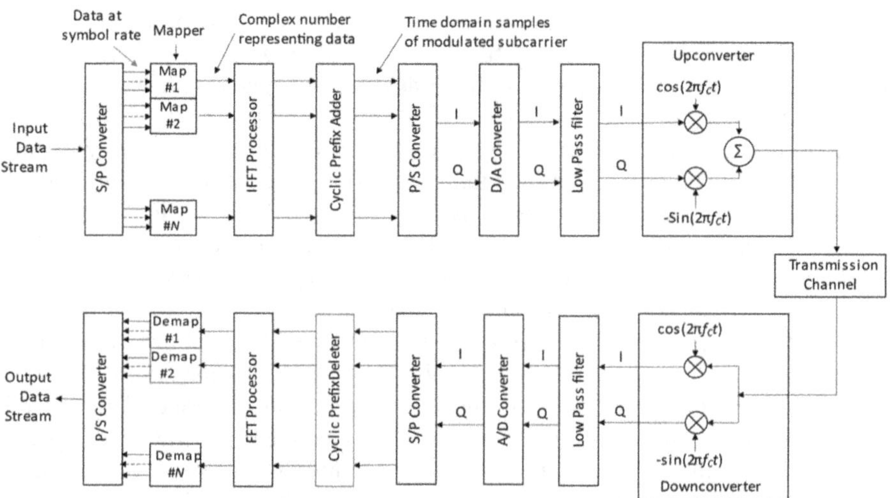

Fig. 6.4   Basic processes in IFFT/FFT-based OFDM system

into the complex number representing the amplitude and phase value of the sub-carrier. For example, if the modulation is 16-QAM and the constellation diagram is as shown in Fig. 4.13b, the 1110 is mapped to the complex number $1 + j3$. The outputs of the mapper feed the IFFT processor. The IFFT knows the unmodulated subcarrier frequency associated with each input and uses this, along with the input, to fully define the modulated signal in the frequency domain. Thus, per symbol period, we have N modulated subcarriers. This set of subcarriers over the symbol period represents one *OFDM symbol* in the frequency domain. An IFFT is performed on this frequency representation, and the output is a set of time domain samples, these samples representing one OFDM symbol in the time domain. The next process shown, the addition of a cyclic prefix, is optional, but almost always applied, and will be discussed below. The outputs of the cyclic prefix adder are fed to the P/S converter where they are broken down into their in-phase (I) and quadra-ture (Q) components and fed out serially in time to create a burst of samples per symbol. Note that the addition of the cyclic prefix can alternately be done on the signal after the P/S converter. The burst of serial samples is then transformed from discrete to analog (continuous) format via the digital-to-analog (D/A) converter and low-pass filter. The output of the low-pass filter is the baseband OFDM signal and is upconverted as shown to create the RF OFDM signal. The upconversion process shown results in a direct transfer of the baseband OFDM signal to RF. At the receiver, the process is reversed. The received waveform is first downcon-verted directly to baseband, low pass filtered, then digitized via an analog-to-digital (A/D) converter. The output of the A/D converter is feed to the S/P converter whose output then feeds the cyclic prefix deleter, should a cyclic prefix have been added in the transmitter. Alternatively, the cyclic prefix deleter can be located immediately following the A/D converter. The next step following the cyclic prefix deleter is conversion back to the complex representation of the symbols by the FFT. These complex representations are then de-mapped to recreate the original parallel data streams, which are then transformed to the original serial stream by the P/S converter.

As indicated above, OFDM is very robust in the face of multipath fading. Nonetheless, in the presence of such fading, a certain amount of ISI is unavoidable unless techniques are implemented to avoid it. Figure 6.5a shows an illustration of how ISI can be incurred as a result of a delayed signal. One technique for eliminating, if not significantly reducing ISI, is the adding of a *guard interval* (GI), or *cyclic prefix* (CP), of length τ_g to the beginning of each subcarrier transmitted useful symbol, of length τ_u, as shown in Fig. 6.5b. This prefix addition allows time for the multipath signals from the previous symbol to die away before the information from the current symbol is processed over the un-extended symbol period. Figure 6.5c is an illustration of how the ISI, shown in Fig. 6.5a, is avoided by using a cyclic prefix. Cyclic prefix addition is carried out, while the symbol is still in the form of IFFT samples and is achieved by copying the last section of the symbol, typically 1/16 to 1/4 of it, and adding it to the front. With this addition, the symbol total duration is now $\tau_t = \tau_g + \tau_u$. Due to the periodic nature of the modulated subcarrier, the junction between the prefix and the start of the original burst is continuous [1]. By adding the guard interval in this manner, the length of the symbol is extended while maintaining orthogonality between subcarriers. As long as the delayed signals from the previous symbol stay within the guard interval, then in the time τ_u there will be no ISI. In the guard interval, there will be ISI, but the guard period is eliminated in the receive process, and the received symbol is processed only over the period τ_u. Should the delayed version of the previous symbol extend beyond the guard interval, however, ISI occurs, but is likely to be limited, as the strength of the delayed signals beyond the guard interval is likely to be small relative to the desired symbol. While adding a prefix eliminates or minimizes ISI, it is not without penalty, as it reduces data throughput, since N symbols are now transmitted over the period τ_t instead of over the shorter period τ_u. For this reason, τ_g is usually limited to no more than about 1/4 of τ_u.

In real OFDM systems, the number of subcarriers actually created, N, is always less than the IFFT processor block size, N_{FFT}. The IFFT process is usually carried out on a total number of samples of size 2^x, where x is a positive integer. In real realizations, therefore, the IFFT processor block size, N_{FFT}, is chosen so that $N_{FFT} = 2^x \geq N$. For example, if number of real subcarriers is 200, then the smallest processor block size or number of "points" would be 256. There would thus, in this example, be 56 "null" carriers. This OFDM signal would be designated as a "256-point FFT" one, even though there would be only 200 real subcarriers. A null subcarrier is a bit of a misnomer as it means a subcarrier location where there is no subcarrier. When present, null subcarriers are usually placed symmetrically above and below the real subcarriers. In addition to being used to carry data, a few subcarriers are often used as pilot subcarriers as alluded to above. Pilot subcarriers are typically used for various synchronization and channel estimation purposes.

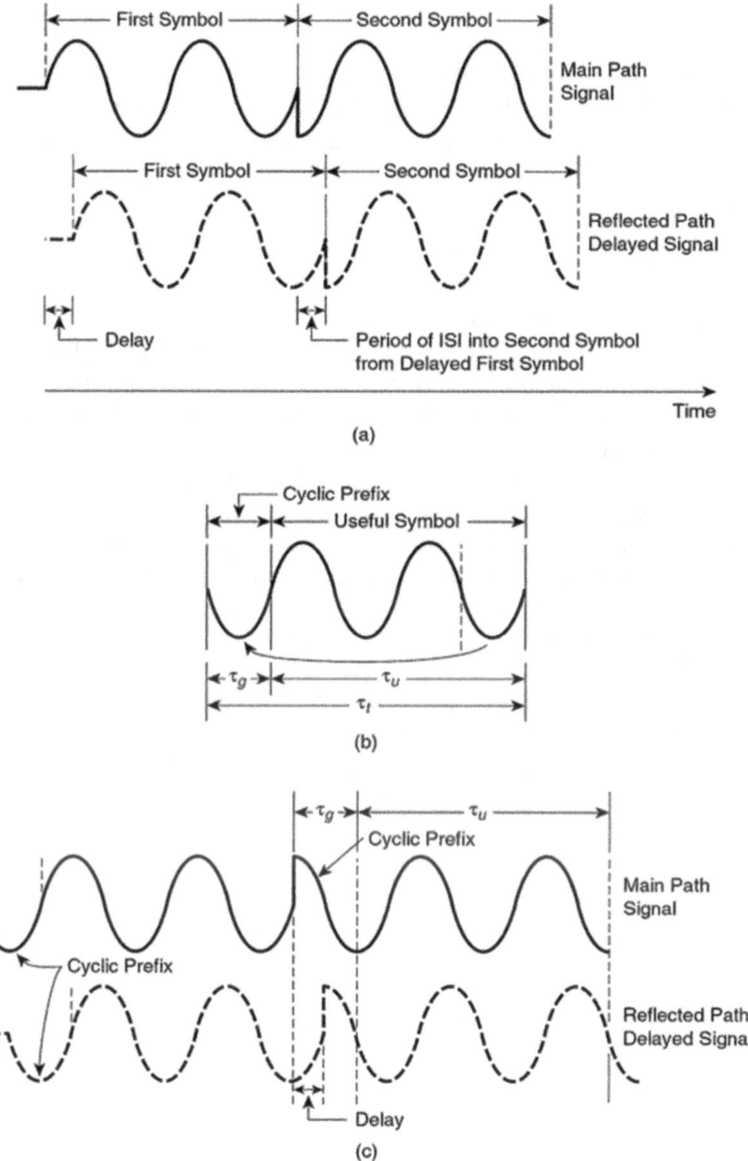

Fig. 6.5 ISI generation and elimination. (**a**) ISI due to delayed signal. (**b**) Cyclic prefix addition. (**c**) Elimination of ISI with cyclic prefix

6.2.2   *Peak-to-Average Power Ratio*

A significant problem with implementation of OFDM is that it exhibits a high *peak-to-average power ratio* (PAPR), where PAPR is defined as the square of the peak amplitude divided by the mean power. The reason for the high PAPR is that in the time domain, an ODFM signal is the sum of N separate subcarriers which are modulated sinusoidal signals. The amplitudes and phases of these sinusoids are uncorrelated, but in the normal course of operation certain input data sequences occur that cause all the sinusoids to add in-phase leading to a signal with a very high peak relative to the average. It can be shown [2] that the probability that the PAPR of an OFDM signal is above a certain threshold, PAPR_0 say, is given by

$$\Pr\left(\text{PAPR} > \text{PAPR}_0\right) = 1 - \left(1 - e^{-\text{PAPR}_0}\right)^N \tag{6.1}$$

where N is the number of subcarriers.

Figure 6.6 shows plots of the distribution of the PAPR given by Eq. 6.1 for different values of the number of subcarriers N. The figure shows that for a given threshold PAPR_0, the probability that the PAPR exceeds that threshold increases with N. Note, however, that this formula implies that the PAPR is independent of modulation order. Numerous computer simulations confirm this.

The input–output characteristics of a power amplifier exhibit a linear range above which nonlinearity sets in until ultimately the output level reaches a maximum. If a high PAPR signal is transmitted through a power amplifier and the peaks of signal fall in the nonlinear region, then signal distortion occurs. This distortion shows up as intermodulation among the subcarriers and out of band emissions. The net result of OFDM signals having a high PAPR is that the power amplifier has to be operated with a large power back off leading to inefficient operation. In addition to potential problems with the transmitter power amplifier, a high PAPR requires a large dynamic range in the transmitter D/A converter and the receiver A/D converter. OFDM PAPR reduction techniques will not be reviewed here but can be found in [3].

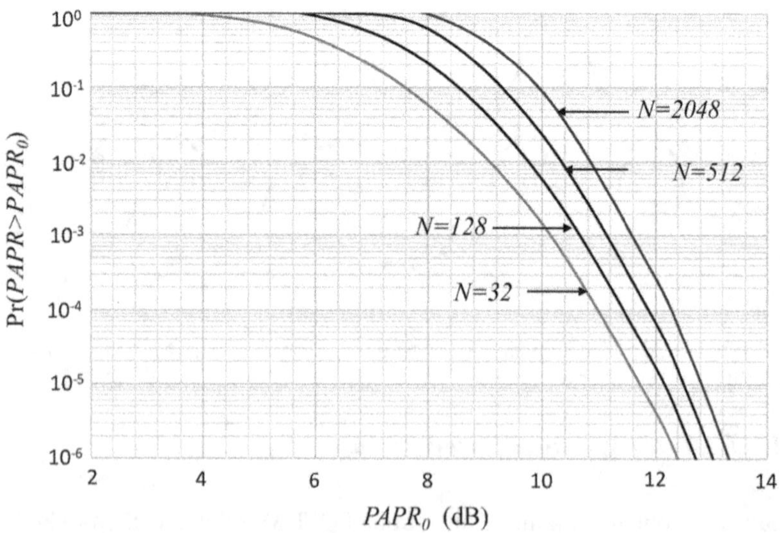

Fig. 6.6 PAPR distribution for different numbers of OFDM subcarriers

6.3   Orthogonal Frequency Division Multiple Access (OFDMA)

OFDM is a multi-carrier transmission technique that enables transmission between two points and where all subcarriers are assigned to one user. However, it lends itself to form the basis of a multiaccess technique for both DL and UL transmission, which is referred to as *orthogonal frequency division multiple access (OFDMA)*. In OFDMA, OFDM subcarriers are divided into sets called *subchannels* and these subchannels assigned to different users in a given time slot, i.e., in a given a number of consecutive OFDM symbols. Subcarriers in each subchannel can be either distributed, i.e., spread over the full channel spectrum available or localized, i.e., structured adjacent to each other. Figure 6.7a,b depicts a three subchannel frequency assignment example where allocation is distributed and localized, respectively.

With distributed allocation, some subcarriers on a given link will likely experience good signal-to-interference and noise ratio (SINR), while others will not, thus experiencing high BER. However, interleaving and coding can minimize the errors generated in the low SINR subcarriers.

Since users typically have different locations, the individual BS/MU links of these users typically suffer from different multipath fading and thus different frequency response across the channel. Localized allocation attempts to exploit this phenomenon by assigning to each user a portion of the available spectrum where channel conditions are good, i.e., with a high SINR. In order to accomplish this, it is obvious that accurate SINR data across the entire useable spectrum must be

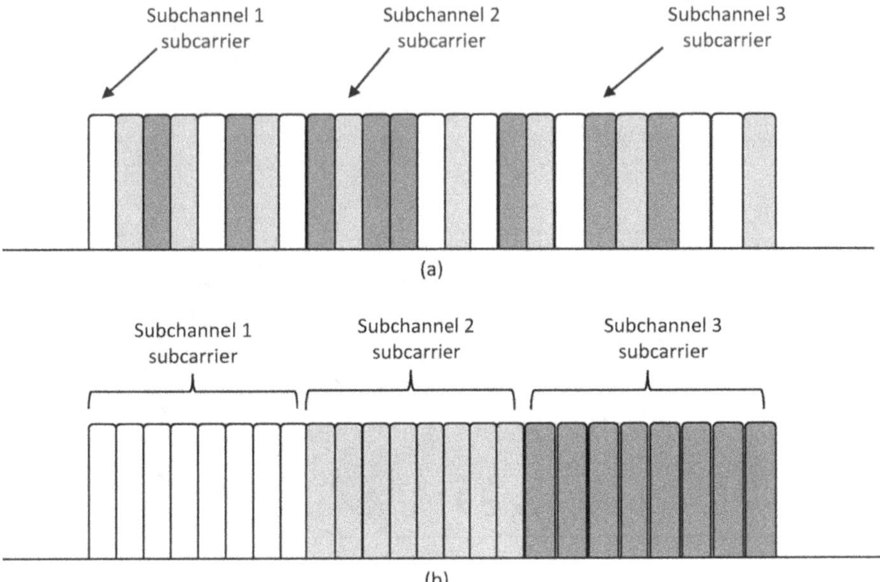

Fig. 6.7 Distributed and localized allocation example. (**a**) Distributed allocation. (**b**) Localized allocation

available to each receiver and that this information be communicated in a timely fashion to its associated transmitter.

If accurate SINR data can be derived at the receiver and applied quickly enough so that a good localized channel can be utilized before the channel condition changes, then localized allocation should outperform distributed allocation. Should this not be the case however, such as where the user is in a highly mobile state, then distributed allocation would be preferred.

With OFDMA as the DL scheme each subchannel is addressed to a different MU, with modulation, coding, etc., tailored to each BS/MU link. With OFDMA as the UL access scheme several MU transmitters may transmit simultaneously with modulation, coding, etc., again tailored to each BS/MU link.

In OFDMA, the scheduler schedules on a two-dimensional (frequency x time) canvas, taking into account the individual user capacity requirements and the channel condition across the full channel bandwidth. In Fig. 6.8, we see an example of how six different users may be scheduled in the time/frequency plane.

OFDMA is used in 5G NR and Wi-Fi 6 in both the downlink and uplink directions. In 5G NR, it is often referred to as cyclic prefix OFDM (CP-OFDM). Localized allocation is used in both systems in both directions.

6.4   Discrete Fourier Transform Spread OFDM (DFTS-OFDM)

Discrete Fourier transform spread OFDM (DFTS-OFDM) is a modified version of OFDMA used optionally by 5G NR in the UL. As in OFDMA, the transmitted signal in a DFTS-OFDM system is a number of orthogonal subcarriers. The PAPR of

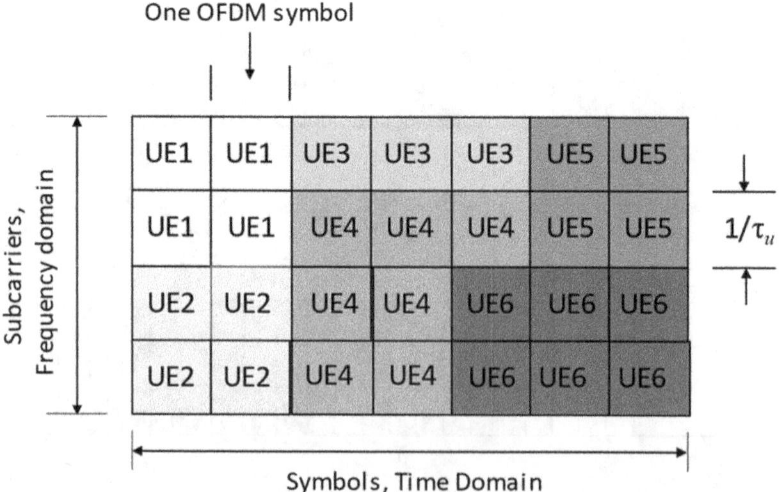

Fig. 6.8 OFDMA scheduling example

DFTS-OFDM is less than that with standard OFDMA. It is this improved PAPR performance of DFTS-OFDM that drives its implementation as this allows higher average transmitted output power and hence higher received signal power.

As was shown earlier, OFDMA takes, per user, N data symbols say (a data symbol being a grouping of input bits), runs them through constellation mappers to create subcarriers in the form of complex numbers, then processes these complex numbers by an M-point ($M > N$) IDFT to generate time domain samples. Thus, within an OFDMA symbol, each subcarrier is modulated with one data symbol. DFTS-OFDM, in contrast, first feeds the N outputs of the mappers into an N-point DFT processor which creates subcarriers in the form of complex numbers and then processes these complex numbers by an M-point IDFT processor to generate time domain samples. The output of the DFT processor spreads input data symbols over all N subcarriers, hence the nomenclature DFT-spread OFDM. The only physical difference between DFTS-OFDM as shown in Fig. 6.9 and OFDMA as shown in Fig. 6.4 is that in DFTS-OFDM, a DFT processor has been added in the transmitter and an IDFT processor added in the receiver.

As with OFDMA, in DFTS-OFDM, there are two defined categories of subcarrier mapping:

- *Localized subcarrier mapping*, referred to as localized FDMA (LFDMA). Here the DFT outputs are allocated to consecutive subcarriers with zeros occupying unused subcarrier positions.
- *Distributed subcarrier mapping*, referred to as distributed FDMA (DFDMA). Here the DFT outputs are allocated over the entire bandwidth with zeros occupying unused subcarrier positions.

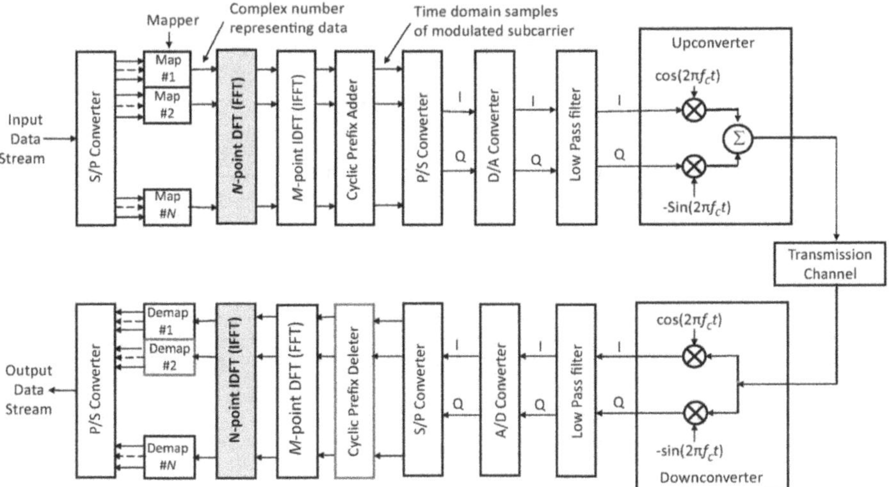

Fig. 6.9 Transmitter/receiver block diagram of DFTS-OFDM

The disadvantage of DFDMA relative to LFDMA is that it provides less flexibility for scheduling. Thus, as with OFDMA, 5G NR employs localized subcarrier mapping with DFTS-OFDM.

6.5   Frequency Hopping Spread Spectrum (FHSS)

In frequency hopping systems, as the name implies, a carrier, modulated with data to be transmitted, is continually changed or "hopped" between a number of predetermined frequencies within a given band so that it spends only a small percentage of its total time at any one frequency. The modulated carrier frequency is pseudo-randomly hopped by controlling the carrier frequency synthesizer with a PN sequence generator. At each frequency hop time the PN generator inputs to the synthesizer a frequency word, of length n chips, which dictates one of $N = 2^n$ frequency outputs. At the receiver, the received signal is de-hopped by downconverting it to an IF frequency with a synthesizer fed by a PN sequence identical to the one used in the transmitter and in alignment with it as received by the receiver. To avoid the need for, and delay associated with, carrier acquisition in the receiver after every hop, M-FSK modulation is normally used in the transmitter and non-coherent demodulation thereof in the receiver. Figure 6.10 shows the block diagram of a typical FHSS radio terminal.

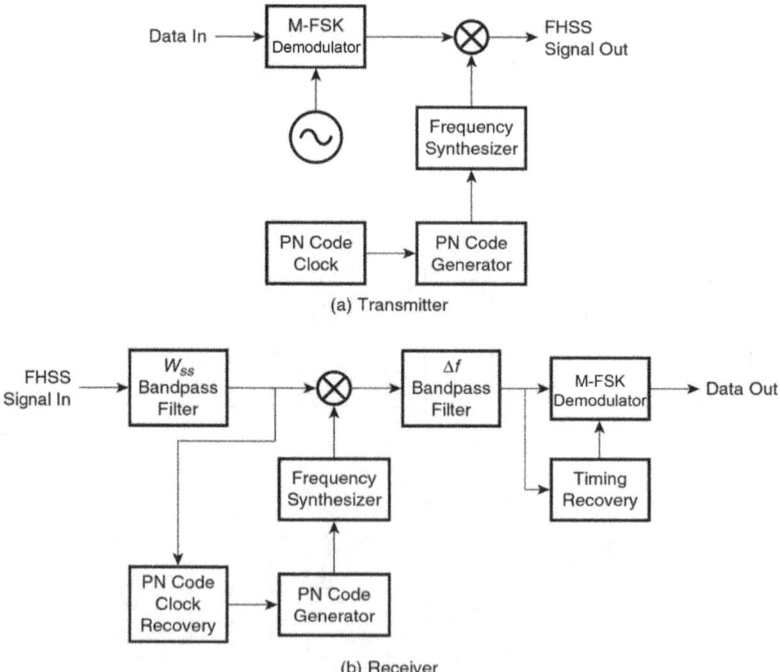

Fig. 6.10   Block diagram of typical FHSS radio terminal

The bandwidth per hop position, Δf, is called the *instantaneous bandwidth*. This bandwidth is that necessary to include most of the power in the M-FSK modulated carrier burst. The bandwidth of the spectrum within which all the hopping occurs, W_{ss}, is called the *total hopping bandwidth* and is equal to $N \times \Delta f$. The processing gain, G_P, of a FHSS system is defined by

$$G_P = \frac{W_{ss}}{\Delta f} = N \tag{6.2}$$

If the hopping rate, also known as the chipping rate, is higher than the transmitted data symbol rate, then there are multiple hops per symbol and the system is referred to as a *fast frequency hopping* one. On the other hand, if the hopping rate is lower than the symbol rate, then there are multiple symbols per hop, and the system is referred to as a *slow frequency hopping* one. The hopping rate of a FHSS is limited by the agility of the transmitter and receiver frequency synthesizers used. The data rate is limited by the data throughput possible in the instantaneous bandwidth. It is further limited by the fact that after each hop the receiver frequency synchronizer and clock recovery circuit need a finite amount of time to lock up, during which time no data can be transferred.

Interference occurs in a FHSS system when an undesired signal occupies a particular hopping channel simultaneously with the desired signal. If the undesired signal is from a non FHSS system with a wide spectrum, the desired signal may encounter the undesired signal at all hopping channels, but the undesired signal's spectral density in each narrow-band hopping channel is likely to be too low to cause significant performance degradation. If the undesired signal is from another FHSS system, however, a *collision* or *hit* is likely to occur, possibly resulting in serious BER degradation, normally in the form of burst errors.

To gain an intuitive understanding of the performance of FHSS system in a FHSS interference environment, consider the case where several users in the same general geographic area independently but synchronously hop their carrier frequencies. Assume that the desired system uses 4-FSK modulation and incorporates no form of error correction. During periods where there are no hits let us assume that the probability of bit error, $P_{be(nh)}$, is given very low. Whenever a hit with a very strong interferer occurs, however, it is reasonable to assume that the output becomes purely random and hence the probability of error, $P_{be(h)}$, is driven to a maximum value of 1/2. Let us assume that $P_{be(nh)}$ is so low that the average probability of error, P_{be}, is controlled only by those generated during hits. Then P_{be} is given by

$$P_{be} = \frac{1}{2} p_h \tag{6.3}$$

where p_h is the probability of a hit.

With N possible hopping channels, the probability that a given interferer will be present in the desired channel, i.e., the probability of a hit, is $1/N$. If there are J

interferers, and assuming N is large, then the probability that at least one interferer is present in the desired slot, i.e., the probability of a hit, p_h, is now J/N. Substituting this value of p_h into Eq. (6.3) we get

$$P_{be} = \frac{1}{2} \cdot \frac{J}{N} \tag{6.4}$$

For example, for a receiver with two high-power interfering signals ($J = 2$) and 37 hopping channels available ($N = 37$), as is the case with Bluetooth LE, then, in an interference environment as described above, P_{be} would equal 2.70×10^{-2}, an unacceptably high performance for most applications. Recall, however, that the above environment assumed no error correction. To minimize the impact of hits, FHSS systems often employ error correction coding that's good at handling burst errors. Another method, often employed if permitted, to minimize the impacts of hits, is to incorporate intelligence into the system that allows it to take note of the location of other users in the band and to adapt its hop sets so as to minimize hits with those other users.

FHSS is the channel usage technique used with Bluetooth. The carrier modulation applied is two-level FSK where the baseband signal is filtered with a Gaussian filter resulting in what is termed Gaussian Frequency Shift Keying (GFSK) which is described in Sect. 4.4.

6.6 Summary

In this chapter, we reviewed the channel usage techniques employed by 5G/5G-Advanced, Wi-Fi 6/7, and Bluetooth 5/6. First, we reviewed orthogonal frequency division multiplexing (OFDM). OFDM is a multi-carrier technique that lends itself to form the basis of multiple-access techniques. One such techniques is employed by both 5G and Wi-Fi, namely orthogonal frequency division multiple access (OFDMA). Another such technique is employed optionally by 5G, namely discrete Fourier transform spread OFDM (DFTS-OFDM). Having reviewed OFDM, we addressed OFDMA and DFTS-OFDM. In the case of Bluetooth, channel usage is via frequency hopping spread spectrum (FHSS). This technique was therefore introduced.

References

1. Morais DH (2021) Key 5G physical; layer technologies, 2nd edn. Springer, Cham, Switzerland
2. Sesia S et al (2011) LTE-the UMTS long term evolution: from theory to practice, 2nd edn. Wiley, West Essex
3. Bisht M, Joshi A (2015) Various techniques to reduce PAPR in OFDM systems: a survey. Int J Signal Process Image Process Pattern Recognition 8(11):195–206

Chapter 7
Multiple Antenna Techniques

Abstract In this chapter, we review at a high level the key multiple antenna techniques applied in 5G/5G-Advanced and Wi-Fi 6/7. Common to all such techniques is the use of multiple antennas at the transmitter, at the receiver, or both, together with intelligent signal processing and coding. These techniques can be broken down into the following three main categories:

- Spatial diversity (SD) multiple antenna techniques: Diversity provides protection against deep fading by combing signals which are unlikely to suffer deep fades simultaneously.
- Spatial multiplexing multiple-input, multiple-output (SM-MIMO) techniques: SM-MIMO permits, in general, the transmission of multiple data streams using the same time/frequency resource thus improving spectral efficiency.

Beamforming multiple antenna techniques: At the transmit end, beamforming permits the focusing of transmitted power in a given direction and thus increasing antenna gain in that direction, and at the receive it permits the focusing of the antenna directivity and hence gain in a given direction.

7.1 Introduction

In this chapter, we review at a high level the key multiple antenna techniques applied in 5G/5G-Advanced and Wi-Fi 6/7. Common to all such techniques is the use of multiple antennas at the transmitter, at the receiver, or both, together with intelligent signal processing and coding. These techniques can be broken down into the following three main categories:

D. H. Morais, *5G/5G-Advanced, Wi-Fi 6/7, and Bluetooth 5/6*, https://doi.org/10.1007/978-3-031-82830-0_7

- Spatial diversity (SD) multiple antenna techniques: Diversity provides protection against deep fading by combing signals which are unlikely to suffer deep fades simultaneously.
- Spatial multiplexing multiple-input, multiple-output (SM-MIMO) techniques: SM-MIMO permits, in general, the transmission of multiple data streams using the same time/frequency resource thus improving spectral efficiency.
- Beamforming multiple antenna techniques: At the transmit end, beamforming permits the focusing of transmitted power in a given direction and thus increasing antenna gain in that direction, and at the receive it permits the focusing of the antenna directivity and hence gain in a given direction.

7.2 Spatial Diversity Multiple Antenna Techniques

Spatial diversity (SD) is enacted at receive end (*receive diversity*) by combining signals from multiple receive antennas and enacted at transmit end (*transmit diversity*) by transmitting signals via multiple antennas. When one transmitter antenna feeds multiple receiver antennas, the system is referred to as a *single-input, multiple-output* (SIMO) one. When multiple transmitter antennas feed one receiver antenna, the system is referred to as *multiple-input, single-output* (MISO) one. Finally, when multiple antennas are employed at both transmit and receive ends to transmit *the same information*, the system is referred to as a *SD multiple-input, multiple-output* (SD-MIMO) one. Figure 7.1 shows the block diagrams of the various configurations of spatial diversity, as well as single-input, single-output (SISO), where there is no diversity.

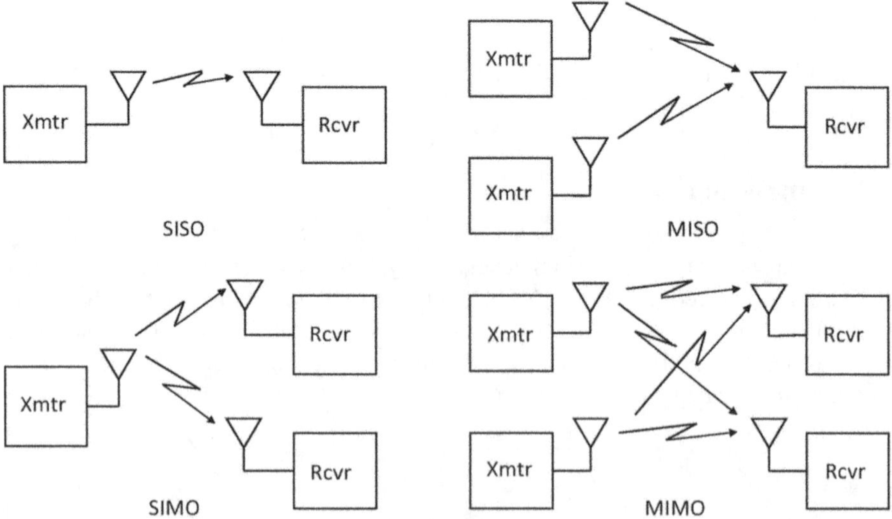

Fig. 7.1 SISO and Spatial diversity configurations

The basic principle behind spatial diversity is that physically separated antennas receive signals or transmit signals that travel over different paths and are thus largely uncorrelated regarding fading. As a result, these signals are unlikely to fade simultaneously. Thus, by carefully combining them, average *signal-to-noise ratio* (SNR) or average *signal-to-interference and noise ratio* (SINR) of an SD system is improved relative to a SISO one. In the case of base station antennas in a typical macro cellular environment, antenna separation on the order of ten wavelengths is typically needed to ensure low mutual fading correlation. For a mobile terminal in a similar environment, however, antenna separation on the order of only half a wavelength is often sufficient to achieve an acceptably low mutual fading correlation, given a large separation at the base station. SD can be used to achieve, with no increase in bandwidth, either of the following:

- Increased reliability via a decrease in average bit or packet error rate.
- Increased coverage area.
- Decreased transmit power.

7.2.1   Space-Time Block Coding

A very effective and one of the most popular forms of MISO spatial diversity systems are ones employing *space-time block coding* (STBC). With STBC the same data is sent via multiple transmit antennas, but each data stream coded differently. At the receiver, STBC algorithms and channel estimation techniques are then used to achieve both diversity and coding gain. In this scheme, the basic idea is to maximize use of both space and time diversity. Information is sent on two or more transmit antennas, with each antenna being fed by its own transmitter. With two transmit antennas, both antennas transmit in the same time slot two different signals, S_1 and S_2 say. They then transmit $-S_2^*$ and S_1^* in the next time slot, where S_x^* is the complex conjugate of S_x, the complex conjugate being created by a simple mathematical manipulation of S_x. Figure 7.2 shows such a system. The net effect is that the transmission rate is the same as if only one transmitter was being used. The reason for swapping signals from one time slot to the next is to statistically diversify the effect of the channel on information and thus increase the chance of correct signal reconstruction. The receiver waits for the received signals contained in the two consecutive time slots and then combines these signals via a few relatively simple computational operations to create estimates of the original signals. Compared to a

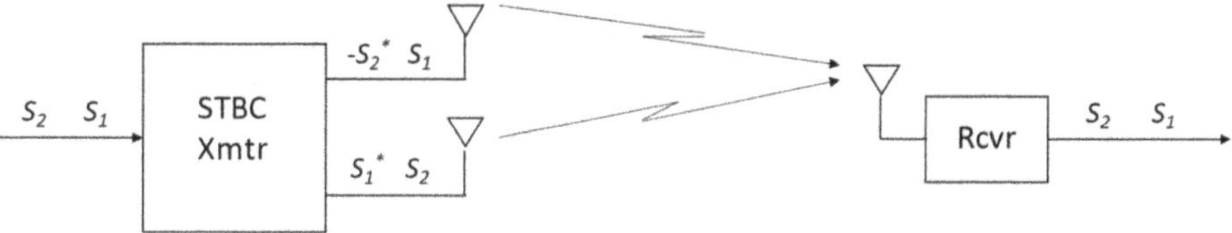

Fig. 7.2 STBC SD-MISO System

SISO system, this scheme reduces the required fade margin by 3 dB under stable conditions, but by more under rapidly fading conditions. It's an option in Wi-Fi 6.

7.3 Spatial Multiplexing MIMO

MIMO systems are very interesting because of their dual capability. They can be used to provide very robust spatial diversity. They can also, however, be used to increase capacity via *spatial multiplexing* (SM), with no additional transmit power or channel bandwidth compared to a SISO system. Such a scheme is referred to as *spatial multiplexing multiple-input, multiple-output* (SM-MIMO). It is important, therefore, to distinguish between SD-MIMO and SM-MIMO. To simplify presentation going forward, "spatial diversity MIMO," when abbreviated, will be indicated as SM-MIMO, while "spatial multiplexing MIMO," when abbreviated, will be indicated as simply MIMO.

7.3.1 MIMO Basic Principles

Consider a MIMO system with N transmit antennas communicating with M receive antennas, where $M \geq N$. At the transmitter, input data is divided via a S/P converter into N substreams. These substreams are then encoded and used to modulate N carriers, each occupying the same channel, and these modulated carriers feed the N antennas. Thus, a "matrix" channel consisting of $N \times M$ spatial dimensions exists within the same assigned bandwidth. Clearly, MIMO increases throughput N times compared to SISO, SIMO, MISO, or SD-MIMO ones. Successful MIMO requires transmission so that the received signals are highly decorrelated. This can be achieved by transmitting the different signals on different polarizations and/or transmitting them over a propagation channel that is rich in multipaths that result from signals bouncing and scattering off nearby objects such as buildings, trees, and cars. At receive end, each of the M antennas picks up all the transmitted substreams and their many images, all superimposed on each other. However, because each substream is launched from a different point in space, each one is scattered slightly differently than the rest. These differences in scattering are key to successful spatial multiplexing transmission. They decorrelate paths taken by the separately transmitted substreams. This minimizes destructive combining at receiver, allowing individual substreams to be identified and recovered via sophisticated signal processing. Recovered substreams are decoded and recombined to recreate original signal. Figure 7.3 shows a capacity-enhancing one-way 2×2 MIMO system.

An analysis of the algorithm used to demultiplex a 3×3 MIMO system can give insight as to how it works. Consider the 3×3 MIMO channel shown in Fig. 7.4 which at the transmit end generates three *spatial streams* (SSs), each intended for a different receiver antenna. Ideally this system results in nine independent transmission channels, each with a scalar coefficient and hence flat fading conditions. Let the coefficient between the receive antenna A_{Ry} and the transmit antenna A_{Tx} and be H_{yx}. The composite signal at each receiving antenna, A_{Ry}, is the addition of the

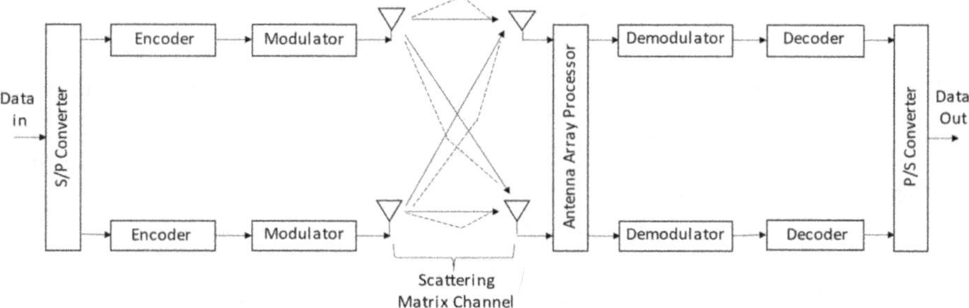

Fig. 7.3 A 2 × 2 MIMO system

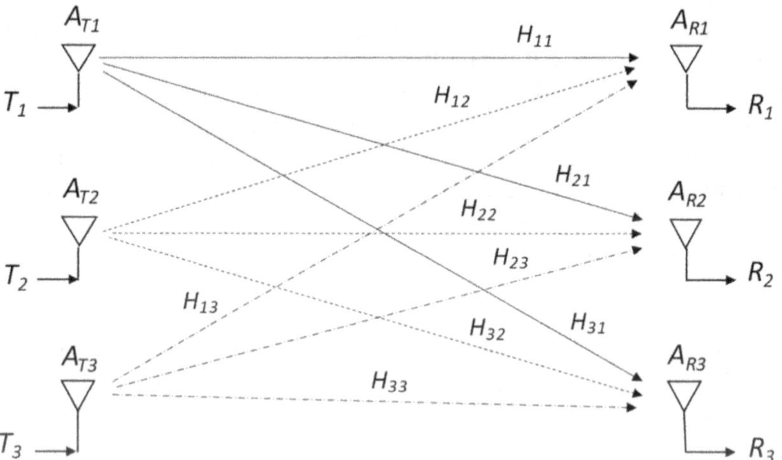

Fig. 7.4 A 3 × 3 SM-MIMO channel

signals received from the three transmit antennas. Thus, for example, $R_1 = T_1 H_{11} + T_2 H_{12} + T_3 H_{13}$. At the receiver, the nine coefficients can be identified through training symbols, achieved, for example, by transmitting a training sequence from one antenna at a time, while the other two transmit antennas are idle. Each transmission allows the determination of three of the coefficients; hence the three independent transmissions allow determination of all nine coefficients. As a result, we have three equations for the three received composite signals and three unknowns, T_1, T_2, and T_3. We are thus able, via high school algebra, to compute the unknowns.

We note that in OFDM systems, decoding is done on a subcarrier by subcarrier basis; hence the assumption in our example of scalar coefficients in the channel matrix is essentially correct, as individual subcarriers are normally sufficiently narrow as to be subjected to flat fading (Sect. 3.2.5.1). Since successful transmission is dependent on multipath scattering, not all remote stations (RSs) will necessarily achieve capacity enhancement. Assuming strong scattering, those close to base station (BS) are likely to be able to achieve it, given high probability here of good SINR. However, those far from BS or those with near line of sight with BS where scattering is weak are unlikely to be able to benefit from capacity enhancement.

7.3.2 Antenna Array Adaptive Beam Shaping

Adaptive beam shaping (*beamforming*) is a technique whereby an array of antennas is used adaptively for reception, transmission, or both in a way that seeks to optimize the transmission over the channel. What we refer to here as beam shaping is also loosely referred to as beamforming, but as beamforming has recently taken on a more specific connotation, the term beam shaping will be used here. In a PMP environment, antenna arrays can be located at both the base station and remote stations (RSs), the latter being referred to as pieces of User Equipment (UE) in 5G NR jargon and Stations (StAs) or clients in Wi-Fi jargon. Omnidirectional or sectorized base station antennas cover an entire cell or sector respectively with almost equal energy, regardless of the location of the remotes. In the case of base station arrays, however, adaptive beam shaping arrays target an individual RS or multiple RSs. It is able to do the latter by creating multiple beams simultaneously, each beam directed to an individual RS. The shape of each beam can be dynamically controlled so that signal strength to and from an RS is maximized, by directing the main lobe in the direction of minimum path loss and side lobes in the direction of multipath components. Furthermore, it can simultaneously be made to minimize interference by signals that arrive at a different direction from the desired by locating nulls in the direction of the interference. Thus, this technique can be made to maximize the SINR. Figure 7.5 shows the beams of an adaptive beam steering antenna array communicating with two remote stations in the presence of multipath and interference.

An *antenna array* consists of two or more individual antenna elements that are arranged in space and interconnected electrically via a feed network in such a fashion as to produce a directional radiation pattern. In adaptive beamforming, the phases and amplitudes of the signals in each branch of the feed network are adaptively combined to optimize the SINR.

Compared to standard omnidirectional or sector antennas at the BS and omnidirectional antennas at the RS, adaptive beamforming antennas, by their ability to focus beams, result in significantly increased coverage and capacity in both LOS and NLOS environments. We note that the gain of a phased array antenna is a direct function of the number of antenna elements. Thus, an antenna array with a large number of elements will be able to transmit a highly focused beam with high gain relative to a single element one. Like the gain, the directivity, i.e., the degree to which the radiated energy is focused in a single direction, is also a direct function of the number of antenna elements.

7.3.3 MIMO Precoding

By application of *precoding* at the transmit end, it is possible to improve MIMO system performance via mild beamforming. Here precoding refers to the linear combining of the original data streams (layers) in the transmitter. When properly applied, this results in an increase and/or equalization of the receiver SINR across

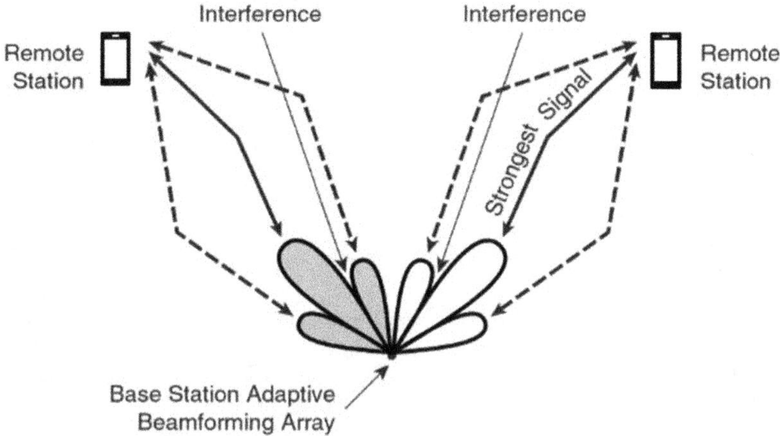

Fig. 7.5 Beams formed from an adaptive beam shaping antenna array communicating with two mobile units

the multiple receive antennas. In a precoded MIMO system, inverse operations are performed at receiver to recover the original un-precoded signals.

Fig. 7.6 shows a simplified version of a 2×2 MIMO system with precoding. On the transmit side of this system, blocks of data bits enter the Modulator which transforms these blocks to corresponding blocks of complex modulation symbols, for example 16-QAM symbols or 64-QAM symbols. The modulator feeds the *Layer Mapper*. The layer mapper creates independent modulation symbol streams (spatial streams), each destined ultimately to an independent antenna. If the mapper is to create n streams, then every nth symbol is mapped to the nth layer. Thus, in our case under study, every second symbol goes to the second layer. The Layer Mapper feeds the Precoder. In the precoder, the layers are combined adaptively via one of a set of weighting matrices. This set of matrices is referred to as a *codebook*. For a 2×2 configuration such as shown in Fig. 7.5, input layers S are multiplied by a weighting matrix W to generate the precoded signals Y given by:

$$\begin{bmatrix} y_1 \\ y_2 \end{bmatrix} = \begin{bmatrix} w_{11} & w_{12} \\ w_{21} & w_{22} \end{bmatrix} \begin{bmatrix} s_1 \\ s_2 \end{bmatrix} \tag{7.1}$$

We note that by matrix multiplication we have

$$y_1 = w_{11}s_1 + w_{12}s_2 \tag{7.2}$$

$$y_2 = w_{21}s_1 + w_{22}s_2 \tag{7.3}$$

The output of the precoder goes to the Resource and Physical Antenna Mapper. Here the modulation symbols to be transmitted on each antenna port are mapped to the set of time/frequency resource elements allocated by the scheduler. By mapping we mean that the modulation symbols modulate specified subcarriers in specified

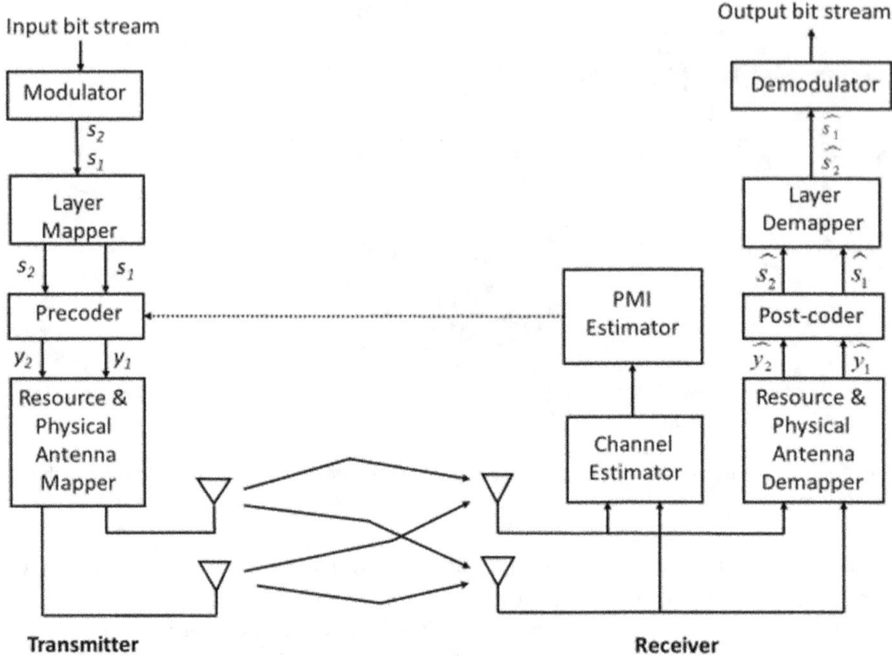

Fig. 7.6 A 2 × 2 MIMO system with a precoding

time slots. Following resource mapping, the modulated subcarriers are mapped to specific antenna ports. On the receive side, the process is reversed. Note, however, that here estimates of the channel matrix H are made via information gleaned from received reference signals on the different antenna ports. These estimates are required to demultiplex the spatially transmitted signals. It is also used, importantly, to feed information back to the transmitter so that the optimum codebook matrix may be applied. From the *singular decomposition* (*SVD*) of H it computes the optimal precoding matrix (SVD description beyond scope of this course). The receiver also knows the codebook. It thus determines which codebook matrix it deems most suitable under current path conditions. The receiver sends this matrix recommendation as a *Precoder Matrix Indicator* (PMI) along with other channel data such as SINR to the transmitter where the final matrix decision is made. Because, in the system described, the receiver feeds back channel derived information to the transmitter, such a system is described as a *codebook based closed-loop precoding* one.

It is also possible to have *non-codebook based transmission*. In the DL, if transmission is via *time division duplexing* (TDD), it operates on one frequency and can only take place in one direction at a time. Hence, the channel is reciprocal, and the BS, rather than relying on a PMI sent by the UE, can generate transmit antenna weights from channel estimates obtained from UL transmissions. The same process can also be used in the UL to allow non-codebook based transmission, but here it's the UE that generates its own transmitter antenna weights.

7.3.4   *Single-User and Multi-User MIMO Operation*

A single-user MIMO (SU-MIMO) system is a spatial multiplexing one where multiple data spatial streams are sent to just one device at a time. It requires multiple antennas at both the transmit and receive end. Its operation means that the full spatial multiplexing gain of the system is directed to one user. The goal of SU-MIMO is to allow the maximization of the individual user data rate. When used with beamforming then the signal is focused on the receive antennas thus increasing the signal-to-noise ratio.

A multi-user form of MIMO, which confusingly is sometimes also referred to as a SU-MIMO, is a spatial multiplexing one where, at any given time, transmission *on a given frequency resource* can only be intended for a given BS/*single RS* link. Note that this does not mean that transmission can only be intended for a single user at any given time, as transmission to other users can be realized via other available frequency resources. Such is the case with OFDMA multi-user access. It can be implemented in both DL and UL utilizing the multiple layers specified. With this form of MIMO the RS thus requires multiple antennas to take advantage of spatial multiplexing. Recall also that for spatial multiplexing between the BS and RS to be effective, the link needs to undergo strong multipath scattering.

A *multi-user MIMO* (MU-MIMO) system, in contrast to the systems described above, is one where, at any given time, transmission *on a given frequency resource* can be between a BS and multiple RSs. Thus, if need be, the full frequency resource can be used for transmission between the BS and each RS. The goal of MU-MIMO is the increase total cell throughput, i.e., capacity. It can operate with only a single antenna at the RS. The system is still a spatial multiplexing one, but here RS antennas are essentially distributed over several RSs. Now, in the DL, if an RS has only one antenna, it does not have to demultiplex several layers and thus the requirement for a channel rich in multipath is removed. In fact, line of sight propagation, which causes considerable performance problems with SU-MIMO, is not a problem here. MU-MIMO can be implemented in both DL and UL.

In DL MU-MIMO, the BS, normally via beam steering, transmits M beams simultaneously to each current user. In a MU-MIMO system that processes signals to/from n RSs, n sets of M weighting elements are used at the BS to produce n outputs. Each set of M weights can null M-1 user signals besides maximizing transmission/reception of a particular desired signal. Thus, for successful multiplexing/demultiplexing of n signals, M must be $\geq n$. MU-MIMO requires that RSs have sufficient angular separation in space as viewed by the BS so that the weighting network can differentiate between them. As a result, MU-MIMO is difficult to implement in mobile environments where RSs may be adequately separated one moment and be in the same direction from BS in the next. MU-MIMO provides a capacity increase as multiple users use the same frequency and time resources simultaneously. Note, however, that for efficient MU-MIMO, the interference between users must be kept low. This can be achieved by using beam shaping as discussed in Sect. 7.3.2 so that when a signal is sent to one user its nulls are formed

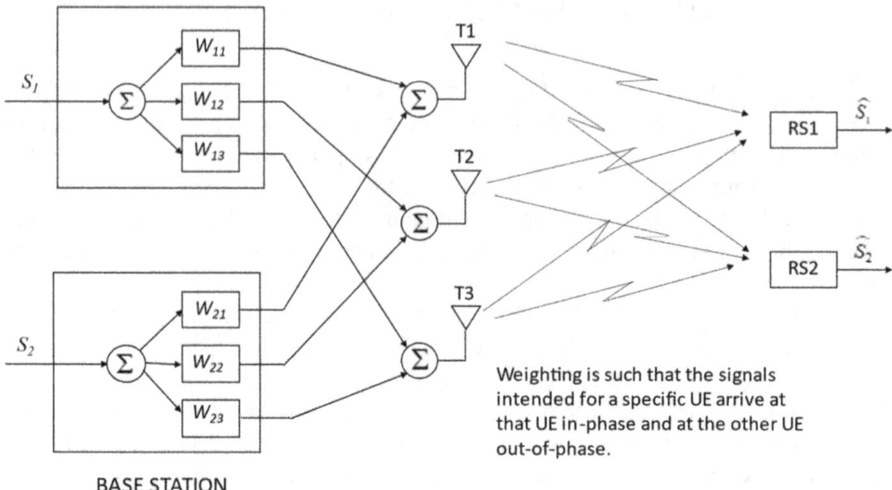

Fig. 7.7 A DL MU-MIMO system showing weighting elements

in the direction of the other users. Figure 7.7 shows a MU-MIMO system where the number of RSs n equals 2 and the number of weighting elements per set, M, and hence number of BS antennas equal 3.

In UL MU-MIMO, also referred to as UL collaborative MIMO, RSs transmit collaboratively in the same time and frequency resource. If say two RSs are transmitting collaboratively, UL throughput is doubled compared to only a single RS transmitting. Note, however, that individual throughput is unchanged. For processing of the received signals, channel estimates are made from known reference signals received. These estimates are then used to determine the weights necessary to cause constructive combination of the signals. Figure 7.8 shows a simplified UL MU-MIMO system.

7.3.5 Massive MIMO and Beamforming

The simplest description of *massive MIMO* (mMIMO) is that it is nothing more than MU-MIMO with a "massive" amount of BS antennas, certainly much more than employed in traditional MU-MIMO, and much more than there are users in the cell. How many BS antennas does it take to create a mMIMO system? There is no definition, but it's generally considered to be about 60 or more. The concept of mMIMO was first proposed in 2010 by Mazetta [1] with the assumption of TDD transmission. As with MU-MIMO, each RS can be allocated the entire bandwidth, and because of the large number of BS antennas, more independent data modulated signals can be sent out, and thus more RSs can be served simultaneously. Broadly speaking, we can think of mMIMO systems as ones with close to a hundred or more of BS antennas serving close to ten terminals or more.

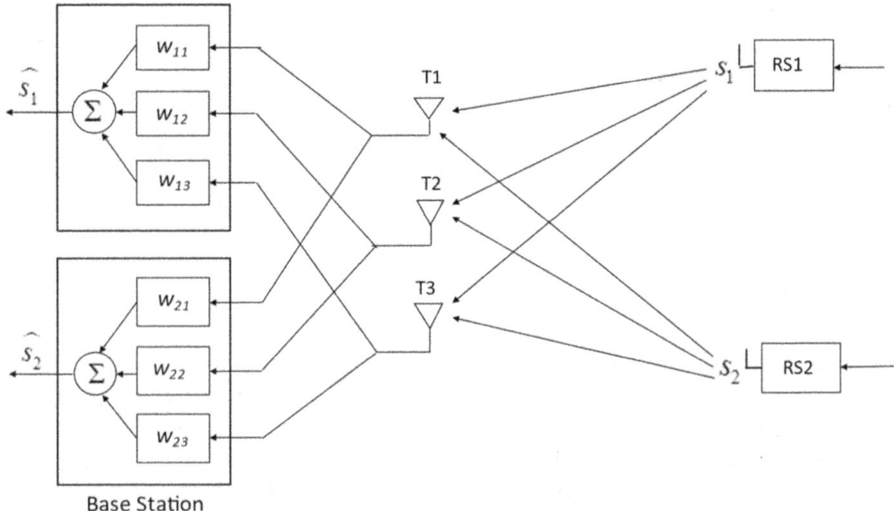

Fig. 7.8 An UL MU-MIMO system showing weighting elements

As with MU-MIMO, the wavefronts emitted from each antenna are weighted so as to add constructively at the intended RS and destructively at other RSs. With mMIMO, because of the large number of antennas directed to the individual RS, coupled with the fact that there are much more BS antennas than likely active users, the overall signal become very directive. Another way of saying this is that the transmitted beam becomes very narrow. This creation of a narrow beam is what is referred to as *beamforming*. A multi-antenna array beamformed signal directed to a single antenna UE can be likened to a laser beam, whereas the more traditional MU-MIMO beam shaping signal can be likened to a flashlight beam. Beamforming is a good technique for situations where there is one dominating propagation path. This is the situation with LOS. It can also, however, be the case where there is no LOS but a strong specular reflection. In the latter situation, the transmitter can focus the beam at the point of reflection and the receiver focus its directivity at the point of reflection.

Some benefits of mMIMO systems are:

- Cell capacity can be increased by at least an order of magnitude. This is because, with increased spatial resolution, more RSs can be served without an unacceptable build up in mutual interference between the RSs.
- Radiated energy efficiency can be improved significantly in both the DL and UL. This is because, with the laser like focus possible, little energy is wasted. In the DL, the effective BS antenna gain is so high that the radiated power per RS can be reduced by an order of magnitude or more. Likewise, in the UL, the effective BS antenna gain is so high that the individual UE transmitted power can be substantially reduced.

- Coverage is improved. Because of the more focused, higher radiated power to each UE, users experience a more uniform, higher data rate service throughout the cell coverage area.
- Features are highly applicable to millimeter wave bands. Propagation characteristics of these bands are generally poor. The high directivity and hence high antenna gain resulting from beamforming can help in overcoming this limitation.

7.3.6 Antenna Array Structure

In mMIMO systems, beamforming is accomplished via *antenna arrays*. These arrays enable high directivity beams and the ability to steer these beams over a range of angles in both the horizontal as well as vertical plane. In general, as was indicated in Sect. 7.3.2 above, the more antenna elements used, the higher the array gain. The beam is steered by controlling the weighting of smaller parts of the array. These smaller parts are called *subarrays*. A subarray typically has multiple antenna elements placed vertically, horizontally, or in a two-dimensional structure. The elements are normally half-wave dipoles, uniformly spaced typically half a wavelength apart, and may either be single polarized or dual polarized. Subarrays are placed side by side to create a uniform rectangular array. Figure 7.9 shows an array with 2 × 1 subarrays, each containing two dual-polarized elements. This array is steered by applying two dedicated radio chains per subarray, one per polarization. It has 16 subarrays and thus 32 transmit/receive RF points of interface.

7.3.7 Full-Dimension MIMO

A MIMO system that allows beam steerability in three dimensions is referred to as a full-dimension MIMO (FD-MIMO) one [2]. This extra degree of freedom allows the more precise directing of a beam to a specific user resulting in higher average

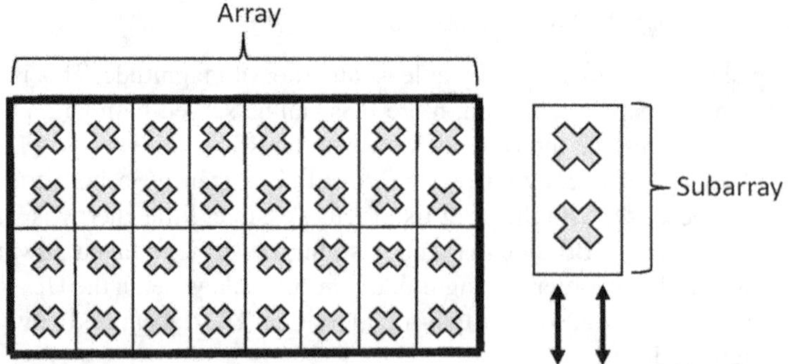

Fig. 7.9 ITU BS antenna array model

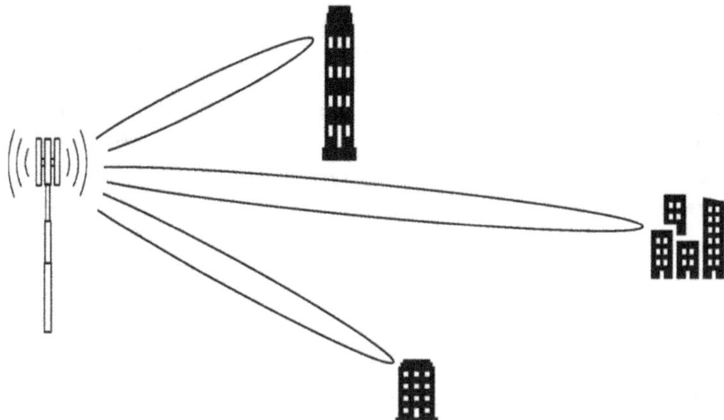

Fig. 7.10 Three-dimensional beamforming

user throughput, less interference at a given UE from beams directed to other UEs, and improved coverage. These benefits of FD-MIMO require narrow targeted beam focusing and thus a large number of array elements. They further require transmission over a LOS path or over one via a strong specular reflection. Thus, to be of value, FD-MIMO is best implemented with mMIMO in the millimeter wave bands.

FD-MIMO is realized by employing *active antenna system* (AAS) technology. With AAS, each array input is integrated with a separate RF transceiver unit. AAS allows the phase and amplitude of the signals to each antenna input to be controlled electronically, resulting in very flexible beamforming. Specifically, it allows multiple beams to be generated, with each individually directable in three dimensions.

In smaller cells such as are found with millimeter wave networks, FD-MIMO is particularly applicable. This is because in such situations the vertical scale is likely to be somewhat comparable with the horizontal one. Figure 7.10 depicts such a situation, very likely to be found in dense urban areas, where three users each require different focusing in all three dimensions. Clearly a lack of control here in the vertical plane would lead to sub optimal coverage.

7.3.8   *Remote Station Antennas*

At sub-6-GHz frequencies, the carrier wavelength is of the same order of the size of the RS and as a result antenna accommodation is typically limited to omnidirectional, dual port, or four port antennas. At millimeter wave frequencies, however, the wavelength is small compared to the size of the RS. This makes it possible to accommodate beamforming antenna arrays with many elements, which is desirable, as this permits relatively large gain that's helpful in overcoming the large path loss. Since the transmission signal can impinge on the UE from any of its sides, it is desirable to have omnidirectional coverage. One way to accomplish this is to have two or more flat arrays that cover different angular sectors, with each array being steerable. With proper placement it should be possible to always have the antenna near optimally directed even when moving.

A complicating factor in RS antenna design is the fact that several millimeter wave frequency bands have to be supported, with the bands differing significantly in frequency. Since array antenna design is wavelength sensitive, many different array designs may need to be incorporated.

7.4　Summary

In this chapter, we reviewed a high level the key multiple antenna techniques applied in 5G/5G-Advanced and Wi-Fi 6/7. Common to all such techniques is the use of multiple antennas at the transmitter, at the receiver, or both, together with intelligent signal processing and coding. Techniques reviewed fell into the following three main categories:

- Spatial diversity (SD) multiple antenna techniques.
- Spatial multiplexing multiple-input, multiple-output (SM-MIMO) techniques.
- Beamforming multiple antenna techniques.

A more comprehensive of discussion of multiple antenna techniques can be found in [3].

References

1. Marzetta TL (2010) Noncooperative cellular wireless with unlimited numbers of base station antennas. IEEE Trans Wirel Commun 9(11):3590–3600
2. Ahmadi S (2019) 5G NR; architecture, technology, implementation, and operation of 3GPP new radio standards. Academic Press, London
3. Morais DH (2021) Key 5G physical layer technologies, 2nd edn. Springer, Cham

Chapter 8
5G/5G-Advanced Overview

8.1 Introduction

In this chapter, key elements of 5G/5G-Advanced are reviewed with particular emphasis on enhanced mobile broadband (eMBB) as this is the feature that most impacts smartphone communication. First, the main architecture options for connection to the core network are reviewed followed by the *Radio Access Network* (RAN) protocol architecture. The RAN is responsible for all radio-related functions such as coding, modulation, HARQ operation, physical transmission, and scheduling. Following this, study within the RAN is narrowed to the physical layer as specified in 3GPP Release 15, the first 5G release, with emphasis on how the user data physical channels are structured. How maximum user data rates and low latency are achieved is demonstrated, 5G operating frequency spectrum reviewed, and some typical base station and UE parameters presented. Next, we address certain key features introduced in Release 16, Release 17, and Release 18 (5G-Advanced).

8.2 Connection to the Core Network

The *core network* is responsible for the overall control of the *user equipment* (UE). Functionally, it sits above the RAN and handles functions not related to radio access but required for the providing of a complete network such as authentication and the establishment, maintenance, and release of communication links. To ease the transition to a fully independent 5G network, 3GPP specified two primary core connection architectures: *non-standalone*, for which there are a number of variations based on the routing of user data, and *standalone*. Figure 8.1 shows the standalone architecture called Solution 2 and one of the non-standalone architectures called Solution 3x.

© The Author(s), under exclusive license to Springer Nature
Switzerland AG 2025

D. H. Morais, *5G/5G-Advanced, Wi-Fi 6/7, and Bluetooth 5/6*,
https://doi.org/10.1007/978-3-031-82830-0_8

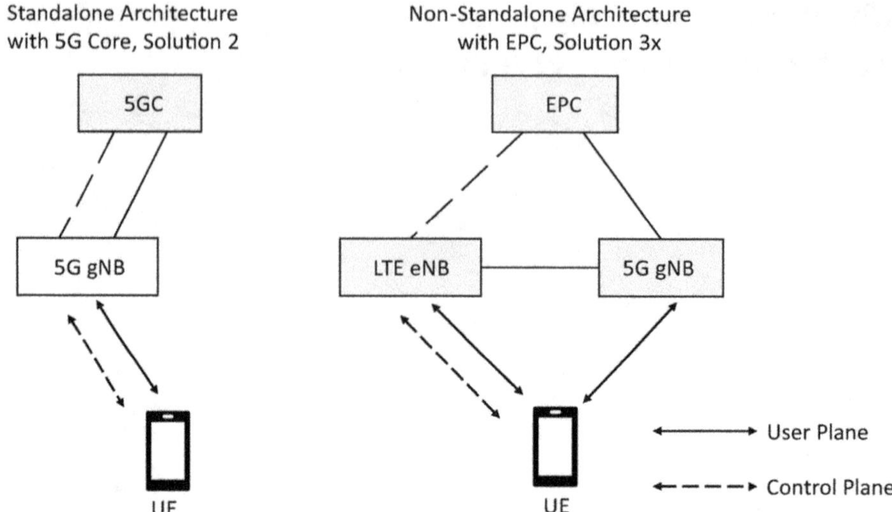

Fig. 8.1 Non-standalone and standalone core connections

In the non-standalone architecture, both the LTE core, called the Evolved Packet Core (EPC), and the LTE base station, called the eNodeB (eNB), are utilized. Control plane data is all done via connection to the eNB, and user plane data to/from the UE is via either the eNB or the NR base station called the gNodeB (gNB) or simultaneously with both. When the latter is the case, this is referred to as *dual connectivity*, which can increase the user data rate and increase reliability. From the point of view of available 5G service, only enhanced mobile broadband (eMBB) is available with the non-standalone architecture. Note, however, that dual connectivity can also be 5G–5G, where the UE communicates with two gNBs.

In the standalone architecture, no elements of LTE are utilized. The UE is connected to the 5G core (5GC) via the gNB and both user plane and control plane data is done via connection to the gNB.

8.3 RAN Protocol Architecture

The 5G RAN protocol architecture of the gNB and the UE [1–3] is shown in Fig. 8.2a. The protocols consist of a *user plane* (UP) and a *control plane* (CP). The UP transports user data which enters/leaves from above in the form of IP packets transported via Ethernet transport links (Sect. 2.2.5). The control plane transports control signaling information and is mainly responsible for connection establishment and maintenance, mobility, and security. The protocol stack is divided into layers. Layer 1 is the physical layer. Layer 2 encompassed the MAC, RLC, PDCP, and SDAP sublayers. Layer 3 is the RRC layer. This layer 3 is specific to the 3GPP RAN architecture and not to be confused with the layer 3 of the TCP/IP protocol stack, namely the Internet Layer protocol described in Sect. 2.2.4. It will be noted

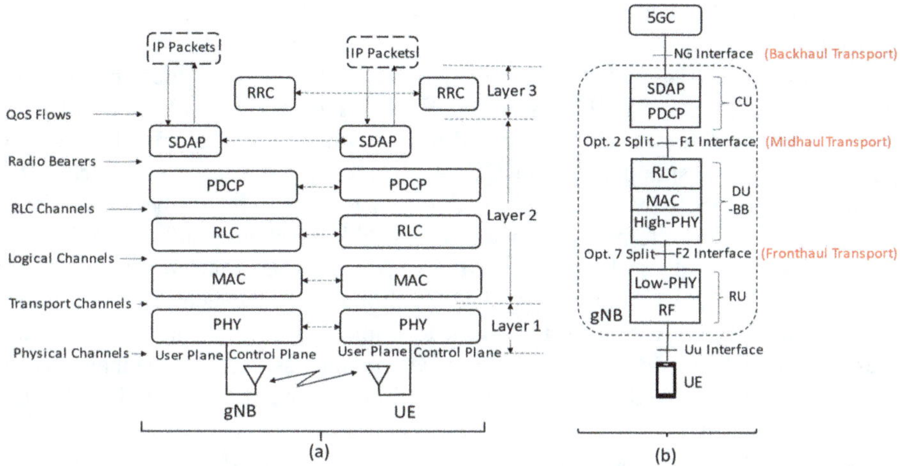

Fig. 8.2 5G architecture. (**a**) RAN user plane and control plane protocols. (**b**) Simplified network architecture

that most of the 5G RAN protocols are common to the user plane and control plane. Shown in Fig. 8.2a is the nomenclature of connections between layer/sublayer and sublayer/sublayer. Logical channels, which are offered to the RLC sublayer by the MAC sublayer, are characterized by the type of information they carry. Transport channels, which are offered to the MAC sublayer by the physical layer, are characterized by how and with what characteristics information is transferred over the radio interface. Data on a transport channel is arranged into *transport blocks* (TBs). Physical channels carry data from higher layers between the gNB and UE.

For our purposes, we note that the 3GPP defined for 5G a RAN architecture where the base station (gNB) functionality is split into two logical units. By logical we mean that a split between functionality is defined but does not necessarily imply a physical split. The two units are the *central unit* (CU) and the *distributed unit* (DU). The CU is connected to the 5GC via the NG interface. The 3GPP studied eight functional splits between the CU and the DU in [4]. It selected Option 2 functional split (PDCP/RLC) for the interface between the CU and DU, this interface being labeled the F1 interface. Within the CU and DU are different layers of functionality. In the DU the lowest layer of functionality is the physical layer which handles among other functions the conversion of data from baseband (BB) format to RF and vice versa. Often the RF functionality is physically separated from the BB functionality. Is convenient therefore to consider the DU as two logical units, namely the DU-BB unit and the *radio unit* (RU) connected via what we shall term the F2 interface, this interface not currently being an officially specified one by the 3GPP. The RU implements the RF functions and may or may not contain some physical layer functions. It interfaces with the user equipment (UE) via the Uu interface. Figure 8.2b shows a highly simplified 3GPP 5G architecture. It shows the CU, DU-BB and RU divisions relative to the 5G NR user plane protocol stack for the specific case where the RU contains some physical layer functions referred to as "Low-PHY" and the DU-BB contains the remaining physical layer functions referred to as "High-PHY." This split corresponds to Option 7 functional split as defined by the 3GPP in [4].

The transport network between the 5GC and the gNB's CU which implements the NG interface, is referred to as *backhaul*. Backhaul payload is IP packet based and so transportable over Ethernet based transport links (Sect. 2.2.5). If there is a physical functional split between the CU and the DU, this arrangement thus calls for a transport network between the CU and the DU which implements the F1 interface, and this network is referred to as *midhaul*. As with backhaul, midhaul payload is packet based and so can be transported over packet-based transport mechanisms such as IP/Ethernet. If there is a physical functional split between the DU-BB and the RU this calls for a transport network to implement the F2 interface between these units. This network is referred to as *fronthaul* and, as with fronthaul and back-haul, is packet based and so can be transported in certain situations via IP/Ethernet transport links.

If the gNB is undivided, the architecture is referred to as a *distributed RAN* (D-RAN) one as the data flow to and from the core network is distributed to each physically undivided gNB. If the CU is separated from the DU the architecture is referred to as a *centralized RAN* (C-RAN) one. Here many of the gNB functions are centralized in the CU which is placed in a more central location enabling more opti-mum network coordination and where one CU can handle a large number of DUs and RF heads.

Before proceeding further, the definition of some commonly used terms in describing the various protocols is in order:

- A *data radio bearer* (DRB) is a radio bearer conveying user data received from the SDAP.
- A *signaling radio bearer* (SRB) is a radio bearer conveying control information received from the RRC.
- A d*ata unit* is the basic unit exchanged between different layers of a proto-col stack.
- A *service data unit* (SDU) is a data unit passed by a layer above to the current layer for transmission using services of the current layer.
- A *protocol data unit* (PDU) is a data unit created by the current layer via the add-ing of a header to the received SDU prior to transportation to the layer below. The header added describes the processing carried out by the current layer.

8.4   Layer 3 (RRC) Description

As shown in Fig. 8.2, Level 3 has one protocol, namely, the *Radio Resource Control* (RRC) protocol, which resides in the control plane. The overall task of the RRC is to configure the UE with the parameters required by the other protocol layers to establish and maintain connectivity between the UE and the gNB.

Note that though referred to here as a Layer 3 protocol, it is not to be confused with the Layer 3 protocol of the TCP/IP protocol stack (Sect. 2.2.4) which is the Internet Protocol (IP).

8.5 Layer 2 User Plane and Control Plane Protocol Description

A detailed layer 2 user plane protocol and control plane structure is shown in Fig. 8.3. Following is a description of the various protocol sublayers, starting from the top.

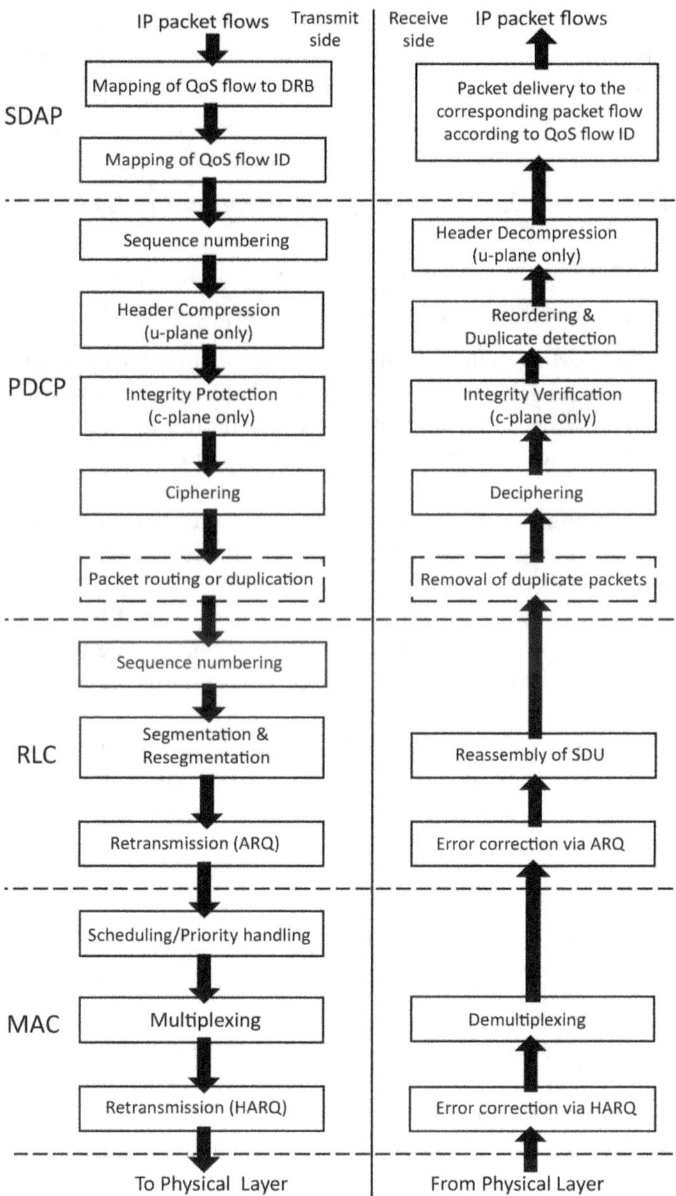

Fig. 8.3 NR user plane and control plane protocol layer structure

8.5.1   *The Service Data Adaptation Protocol (SDAP) Sublayer*

In NR, IP packets as described in Sect. 2.2.4 are mapped from the 5G core network to data radio bearers (DRBs) according to their quality of service (QoS) requirements. The *Service Data Adaptation Protocol* (SDAP) is a user plane sublayer and is responsible for this mapping. It also marks both the DL and UL packets with an identifier as to the specified QoS. The SDAP is a new protocol and not present in 4G. If the gNB is connected to the EPC, as in the non-standalone case shown in Fig. 8.1, the SDAP is not used.

8.5.2   *The Packet Data Convergence Protocol (PDCP) Sublayer*

The *Packet Data Convergence Protocol* (PDCP) is responsible for several services and functions [1]. The following is a description of the main ones:

- Sequence numbering, in the downward flow in the user plane, of data radio bearers so that in-sequence reordering is possible in the upward flow.
- Header compression, in the user plane only, to reduce the number of bits transmitted over the radio interface. In the upward flow, header decompression is applied.
- Integrity protection, in the downward flow in the control plane, to ensure that control messages originate from the correct source. In the upward flow, integrity verification is applied.
- Ciphering, also known as encryption, in the downward flow in both the user plane and control plane, to ensure that intruders cannot access the data and signaling messages that the UE and the gNB exchange. In the upward flow, deciphering is applied.
- Duplicate detection and removal in the upward flow in the user plane and, if in order delivery to layers above is required, reordering to provide in-sequence delivery.

8.5.3   *The Radio Link Control (RLC) Sublayer*

The RLC sits below the PDCP sublayer and interfaces with the MAC sublayer below via logical channels. Depending of the class of service to be provided, the *Radio Link Control* (RLC) sublayer can be structured in one of three transmission modes, namely, the *transparent mode* (TM), the *unacknowledged mode* (UM), and the *acknowledged mode* (AM). In the transparent mode, data unit flow is transparent, and no headers are added. In the unacknowledged mode, segmentation (discussed below) is supported, and, in the acknowledged mode, segmentation, and retransmission of erroneous packets are supported. The RLC is responsible for several functions and services including:

- Sequence numbering of user data in the downward flow, independent of the one in the PDCP above. This numbering is in support of HARQ retransmissions. In the upward flow, RLC does not facilitate in-sequence delivery, minimizing the overall latency. This task is left to the PDCP sublayer above.
- Segmentation in the downward flow, of RLC SDUs, received from the PDCP above, into suitably sized RLC PDUs. In the upward flow, reassembly of the original RLC SDUs.
- Retransmission, in the acknowledgment mode, in the downward flow, via HARQ, of erroneously received RLC PDUs at the receiving end.
- Error correction in the upward flow via HARQ.

8.5.4   The Medium Access Control (MAC) Sublayer

The *Medium Access Control* (MAC) sublayer is the lowest sublayer of Level 2 and interfaces with the physical layer below via transport channels. As indicated above, data on a transport channel is arranged into transport blocks (TBs), and the transmission time of each transport block is called the *Transmission Time Interval* (TTI). One transport block of variable size can be transmitted in each TTI over the radio interface to/from an EU except when there is spatial multiplexing of more than four layers, in which case two transport blocks are transmitted per TTI.

The MAC sublayer is responsible for several functions and services including:

- Mapping between logical channels and transport channels.
- In the downward flow, multiplexing of MAC SDUs (RLC PDUs) belonging to one or different logical channels into transport blocks delivered to the physical layer on transport channels.
- In the upward flow, demultiplexing of MAC SDUs belonging to one or different logical channels from transport blocks delivered from the physical layer on transport channels.
- Error correction via HARQ.
- Scheduling and scheduling-associated functions.

The MAC sublayer provides services to the RLC sublayer above via logical channels that can be classified into two groups: traffic channels used for carrying user plane data and control channels used for transporting control plane signaling. Logical channels are mapped to transport channels and transport channels in turn are mapped to physical channels.

8.5.5   *Layer 2 User Plane Downward Data Flow*

An example of Layer 2 user plane downward data flow is shown in Fig. 8.4. Here, at the top, are three IP packets, two on radio bearer RB_x and one on radio bearer RB_y. First the IP packets enter the SDAP layer as SDAP SDUs where they each have a header added to create SDAP PDUs. These PDUs enter the PDCP layer as PDCP SDUs where, after processing including header compression, each has a header added to create PDCP PDUs. These PDUs enter the RLC layer where the ones created from radio bearer RB_x become RLC SDUs and the one created from radio bearer RB_y is segmented into two RLC SDU segments. After processing, the segmented and unsegmented SDUs each have a header added to create RLC PDUs which are then forwarded to the MAC layer where they are received as MAC SDUs. These SDUs, after processing, each have a header added creating MAC PDUs. The first three MAC PUDs counting from left to right (two from RB_x and one from RB_y) are multiplexed to form one transport block which is forwarded to the physical layer, and the fourth MAC PDU (from RB_y), on its own, is forwarded to the physical layer as another transport block.

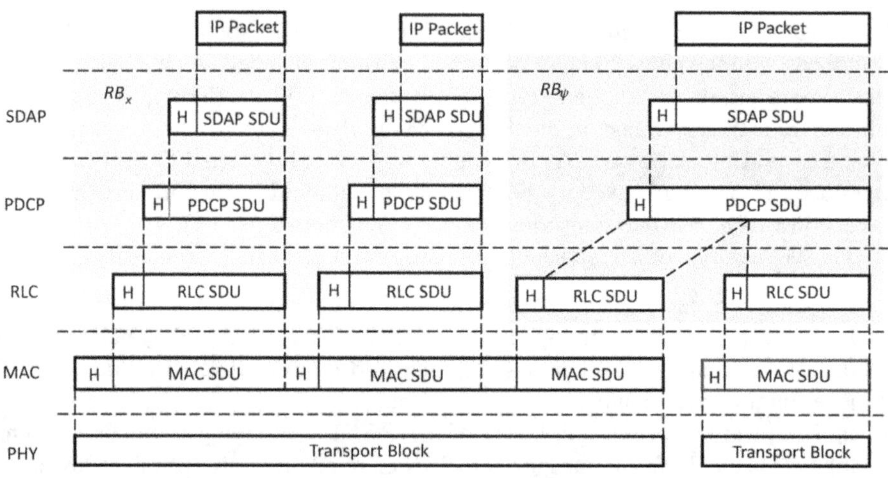

Fig. 8.4 An example of Layer 2 user plane data flow

8.6   Layer 1 (Physical Layer)

The physical layer is the lowest layer, Layer 1, in the RAN protocol architecture. Its input on the downward flow and output on the upward flow is transport blocks. As mentioned above, one transport block of variable size can be transmitted in each TTI over the radio interface to/from an EU except when there is spatial multiplexing of more than four layers, in which case two transport blocks are transmitted per TTI.

The physical layer is responsible for many functions including CRC attachment, coding, rate matching and HARQ, scrambling, linear modulation, layer mapping, and mapping of the signal to the assigned physical time–frequency resource. It also handles the mapping of transport channels to physical channels. For the transportation of user data two physical channels are defined in NR specifications, namely the *Physical Downlink Shared Channel* (PDSCH) and the *Physical Uplink Shared Channel* (PUSCH).

8.6.1   Numerologies

NR uses OFDM-based transmission in both the UL and DL and there are a number of specification options associated this transmission. Here we examine what is referred to as *numerology*. A numerology (μ) is defined by *subcarrier spacing* (SCS) and cyclic prefix (CP) overhead. In LTE the subcarrier spacing is 15 kHz except for one special case (multicast broadcast single frequency network) where it is 7.5 kHz. In NR several numerologies are supported which can be mixed and used simultaneously. Multiple subcarrier spacings are derived by scaling a basic subcarrier spacing (15 kHz) by an integer μ. The subcarrier spacing Δf for numerology μ is given by $\Delta f = 15 \times 2^{\mu}$ kHz. The numerology applied is selectable, independent of the frequency band, but for reasons that will be discussed below, a low subcarrier spacing is not recommended for very high carrier frequencies. Table 8.1 provides supported numerologies in NR along with associated subcarrier spacing, OFDM useful

Table 8.1 NR numerology

Numerology	Subcarrier spacing (kHz)	OFDM useful symbol length (µs)	Cyclic prefix duration (µs)	OFDM symbol length (µs)
0	15	66.67	4.69	71.35
1	30	33.33	2.34	35.68
2	60	16.67	1.17	17.84
2	60	16.67	4.17 (extended)	20.84
3	120	8.33	0.59	8.92
4	240	4.17	0.29	4.46
5	480	2.08	0.15	2.23
6	960	1.04	0.07	1.11

symbol length, cyclic prefix duration, and OFDM symbol length. We note that numerologies 0–4 were supported in 3GPP Release 15, and numerologies 5 and 6 were added in Release 17. We further note that numerology 4, with a subcarrier spacing of 240 kHz, is not supported for data channels, being only supported for what is referred to as synchronization signal block transmission in the FR2 range.

There are many advantages to offering multiple SCS options. Symbol length is inversely proportional to SCS. Thus, with the smaller SCSs, we have the larger symbol lengths. As NR uses the same ratio of CP length to OFDM symbol length, the smaller SCSs have the larger CP lengths and are thus the most tolerable to the effects of multipath delay spread. With the larger SCSs, on the other hand, the symbol durations are shorter. This results in faster transmission turnaround. Also, it results in lower sensitivity to phase noise, since the negative impact of phase noise reduces with subcarrier spacing. Finally, it minimizes the possible degrading effect of Doppler shift (Sect. 3.2.6.4).

8.6.2   The Frame (Time Domain) Structure

The NR frame structure supports both frequency division duplexing (FDD) and time division duplexing (TDD) operations (Sect. 8.6.4.5) in both licensed and unlicensed frequency bands. It facilitates very low latency, fast acknowledgment of HARQ, and dynamic TDD. Transmissions in the time domain, in both the DL and UL, are organized into *frames*. There is one set of frames in the UL and one in the DL on a carrier. Each UL frame transmitted from the UE starts at a specified time before the start of its corresponding DL frame at the gNB to ensure that the UL frame, when it arrives at the gNB, is synchronized with the DL frame at the gNB.

Frames are of 10-ms duration, each frame being divided into ten *subframes*, with each subframe thus being of 1-ms duration. In turn, each subframe is divided into *slots*, each slot consisting of 14 OFDM symbols. A slot represents the nominal minimum scheduling interval, referred to as the Transmission Time Interval (TTI), but data can be scheduled to span more than one slot, the latter being referred to as slot aggregation. Recall from Table 8.1 that the OFDM symbol varies with numerology. Thus, the duration of a slot and hence the number of slots per subframe depend on the numerology. This is demonstrated in Fig. 8.5 [5]. The larger the subcarrier spacing, the shorter the slot duration. For subcarrier spacings of 15, 30, 60, and 120 kHz, the associated slot durations are 1, 0.5. 0.25, and 0.125 ms, respectively. In terms of numerology, slot duration is equal to $1/2^{\mu}$ ms. The shorter the slot duration, the lower the latency. This lowered latency is achieved, however, at the expense of a shorter cyclic period which, in a large cell deployment, may not be acceptable due to large rms delay spread. To further aid in latency reduction, reference signals and control signals are located at the beginning of the slot or the set of slots in the case of slot aggregation, as this speeds up the receiver processing.

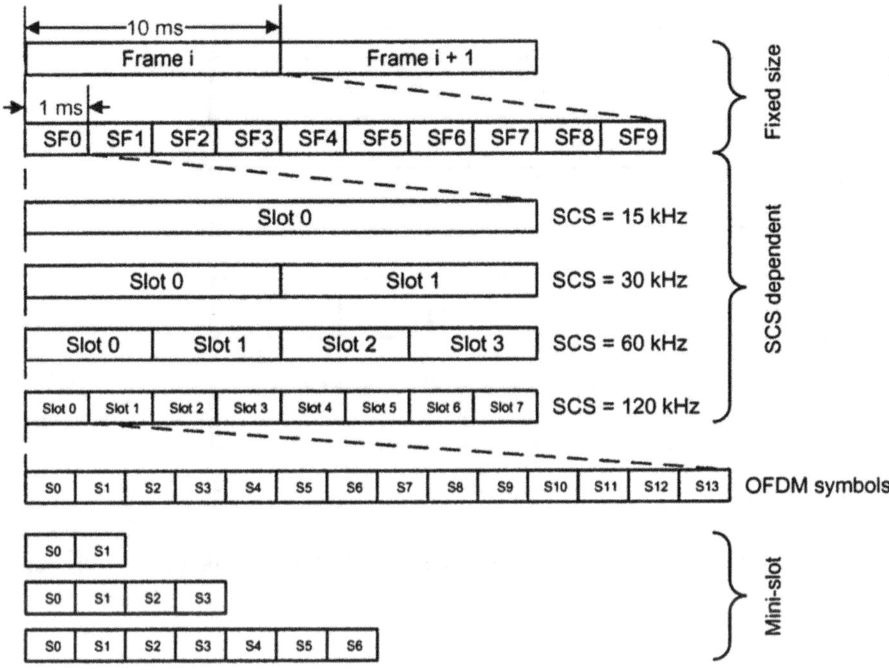

Fig. 8.5 Frame structure and slots. (From [4], with the permission of Elsevier)

To allow greater flexibility regarding latency, NR allows using less than 14 symbols for transmission, the resulting smaller effective slots being referred to as *mini-slots*. With the mini-slot structure, 2, 4, or 7 symbols can be allocated with a flexible start position allowing transmission to commence as soon as possible without waiting for the start of a slot boundary. Mini-slots permit shorter latencies even with the 15-kHz subcarrier spacing. There are three good reasons for allowing the use of mini-slots. The first is lowered latency in lower frequency transmissions where typically 15- or 30-kHz subcarriers are employed. The second is support of analog beamforming in very high-frequency transmission. Here, even though high subcarrier frequencies are used and hence shorter slots, only one beam at a time can be used for transmission and thus transmission to multiple users must be on a time division basis. Without mini-slots, all 14 symbols would have to be used per UE, before moving on to the next UE. The net effect would be increased multi-transmission latency. With the use of mini-slots, this can be reduced even when transmitting large payloads as with the very large bandwidths available, a few OFDM symbols can be of large data-carrying capacity. The third good reason for employing mini-slots is to facilitate efficient transmission in unlicensed bands. In such bands, the transmitter must constantly monitor the band usage and only transmit when the band is clear. Once it is determined that the band is clear, then the quicker transmission can begin, the better the chance of seizing the spectrum. Mini-slots facilitate this quick action.

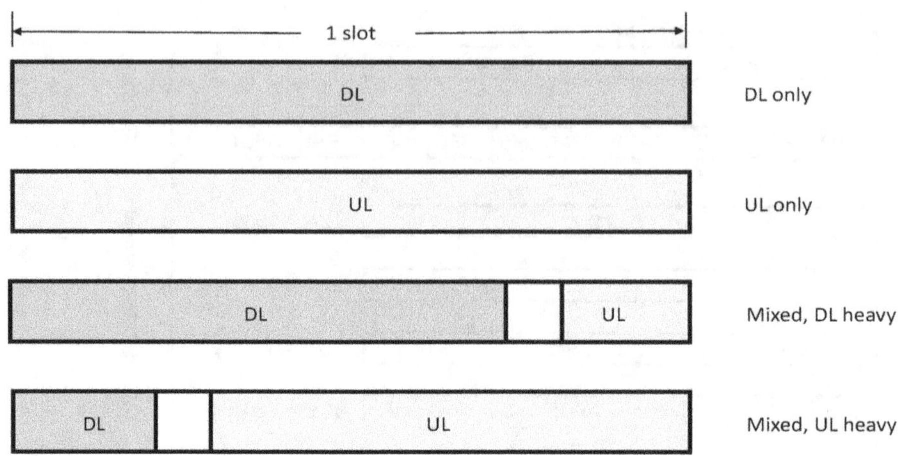

Fig. 8.6 TDD-based frame structure

In the TDD mode, as shown in Fig. 8.6, a slot can be scheduled for all DL transmission, all UL transmission, or a mixture of both DL and UL, where guard periods are inserted for UL/DL switching.

8.6.3   The Frequency Domain Structure

In NR, a *resource element*, consisting of one subcarrier during one OFDM symbol, is the smallest physical resource specified. In the frequency domain, 12 consecutive subcarriers of the same spacing are called a *resource block* (RB), the width of a resource block thus being a function of its numerology. The basic scheduling unit is the *physical resource block* (PRB), which is defined as 12 consecutive subcarriers of the same spacing in the frequency domain, i.e., 1 RB, over 1 OFDM symbol in the time domain, with all subcarriers within the PRB having the same CP length. Multiple numerologies are supported on the same carrier, and resource block locations are specified so as to have their boundaries aligned. Thus, as shown in the PRB alignment portion of Fig. 8.7, 2 PRBs of subcarrier spacing 30 kHz, for example, occupy the identical frequency range as 1 PRB of 60-kHz subcarriers. Resource grid portion shown in Fig. 8.7 presents a resource element (dark shading) and a PRB (light shading) within a physical resource grid, the latter being described below.

For NR, the IFFT processor block size, N_{FFT}, has been set at 4096 (2^{12}). Thus, in theory, up to 4096 subcarriers could be accommodated per NR carrier. However, as there are always null subcarriers at the low and high ends of the channel, 3GPP has set the maximum number of subcarriers per NR carrier at 3300. As there are 12 subcarriers per PRB, this leads to a maximum number of PRBs per NR carrier of 275.

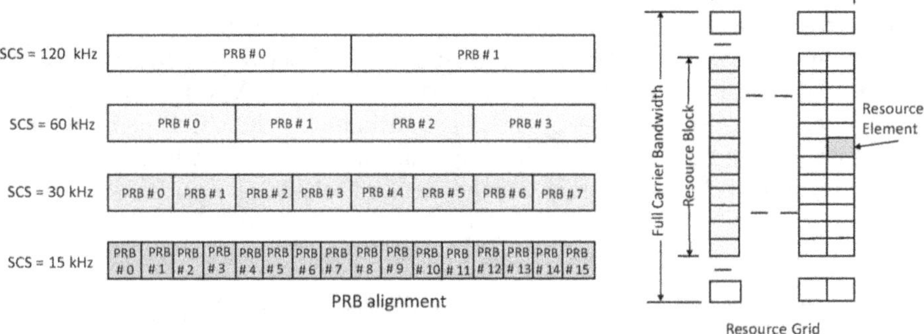

Fig. 8.7 PRB alignment and resource grid

Table 8.2 Minimum and maximum number of resource blocks and corresponding occupied transmission/channel bandwidths

Freq. range	Subcarrier spacing (kHz)	Min (N_{PRB})	Max (N_{PRB})	Minimum transmission/ channel bandwidth (MHz)	Maximum transmission/channel bandwidth (MHz)
FR1	15	25	270	4.50/5	48.60/50
FR1	30	11	273	3.96/5	98.20/100
FR1	60	11	135	7.92/10	97.20/100
FR2-1	60	66	264	47.52/50	190.08/200
FR2-1	120	32	264	46.08/50	380.16/400
FR2-2	120	66	264	95.04/100	380.16/400
FR2-2	480	66	248	380.16/400	1428.48/1600
FR2-2	960	33	148	380.16/400	1704.96/2000

NR specifies minimum and maximum values of PRBs (same for DL and UL) per carrier as a function of numerology. Table 8.2 provides these values and their corresponding transmission bandwidths for frequency ranges FR1, FR2-1, and FR2-2 (Sect. 8.8). Note that the transmission bandwidth is the actual bandwidth occupied by the resource blocks, whereas the channel bandwidth is the bandwidth assigned to a single NR RF carrier and includes guard bands. For each numerology and carrier, a *resource grid* is defined that covers, in the frequency domain, the full carrier bandwidth being utilized and, in the time domain, one subframe. The resource grids for all subcarrier spacings overlap, and there is one set of resource grids per transmission direction. The resource grid defines the transmitted signal space as seen by the UE for a given subcarrier spacing. However, the UE needs to know where exactly in the available transmitted bandwidth its associated resource blocks are located. To address this need, NR specifies a common reference point for resource grids referred to as *point A* as well as two classes of resource blocks, namely, *common resource blocks* and *physical resource blocks*. Common resource blocks are numbered from

0 and upward in the frequency domain for a given subcarrier spacing. The center of subcarrier 0 of common resource block 0 for a given subcarrier spacing coincides with "point A." Point A serves as a reference from which the frequency structure can be described and need not be the actual carrier frequency. As a part of the initial access procedure, the location of point A is transmitted to the UE as part the information broadcasted by the PBCH. The physical resource blocks, which indicate the actual transmitted signal spectrum, are then located relative to point A.

8.6.4   Physical DL Shared Channel (PDSCH) and Physical UL Shared Channel (PUSCH) Processing

The DPSCH and the PUDSCH are the physical channels that carry user data. We will thus study these channels in some detail as this will help us to understand how NR can achieve very high data rates. Figure 8.8 shows a block diagram summary of the various processing steps taken on the transmit side by the user data-carrying PDSCH and PUSCH from transport block(s) into the gNB or UE to antenna out (up-conversion is omitted). In the receiver, the processing is the inverse of that in the transmitter. The overall data processing in these channels is largely similar. Most of the key technologies required in this processing have already been covered in prior chapters of this text, and, where so, the appropriate section is indicated on the figure. The purpose of this section is to cover those necessary technologies not yet covered (shown shaded), some simple but key to the process chain, and to the expand discussion on selected ones previously covered. Thus, by the end of this section, we will have covered the key technologies necessary to comprehend in some detail 5G NR physical layer processing of the main data-carrying channels.

8.6.4.1   LDPC Base Graph Selection and Code-Block Segmentation

In NR, one or two data-carrying transport block(s) of variable size is delivered to the physical layer from the layer above, the Medium Access Control (MAC) layer. The first action at the physical layer is the attachment, per transport block, of a CRC (Sect. 5.2), CRC_{TB} say, as shown in Fig. 8.9, thus creating a *code block*. Next, the LDPC base graph to be used is selected based on the transport block size and the coding rate indicated (Sect. 5.3.4). Code-block segmentation follows. Here, should a transport block exceed a certain size (8424 bits for LDPC base graph 1, 3824 for base graph 2), the transport block and its attached CRC is broken up into several equal sized code blocks, and each code block has a CRC, CRC_{CB} say, of length 24 bits, attached as shown in Fig. 8.9.

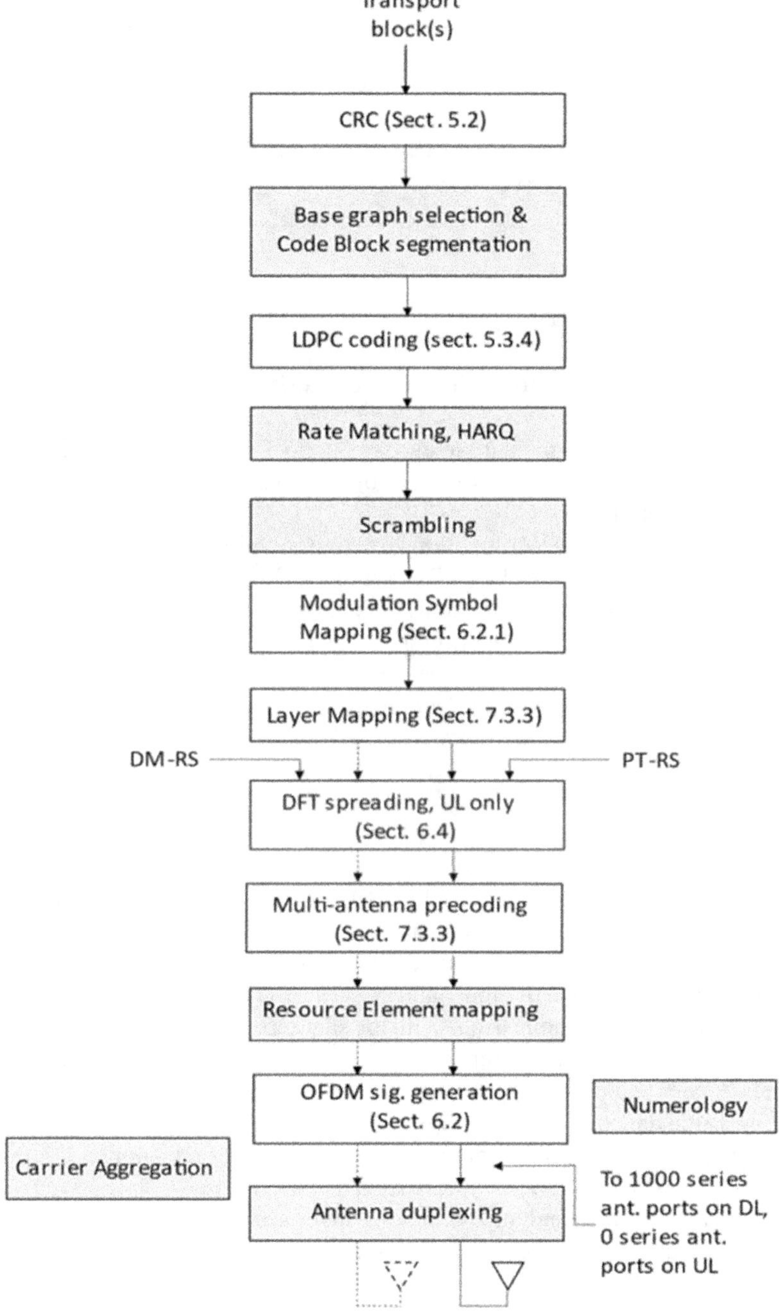

Fig. 8.8 Physical layer processing

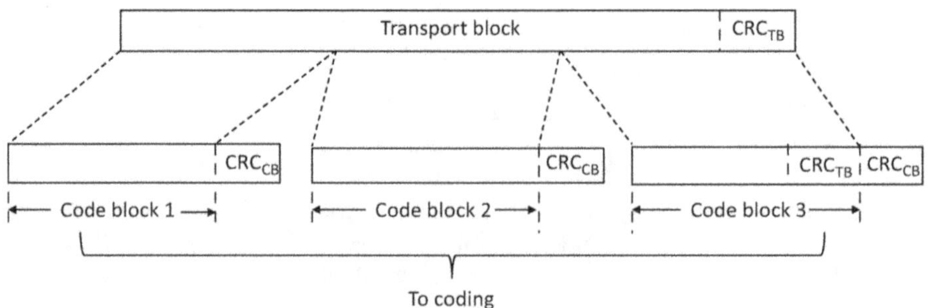

Fig. 8.9 Code-block segmentation

8.6.4.2 Rate Matching and HARQ Functionality

Following code-block segmentation, each code block and its attached CRC undergoes LDPC coding (Sect. 5.3.4), followed by rate matching and HARQ functionality [2], the subject of this section.

Rate matching and HARQ functionality, performed separately for each LDPC coded code block and its attached CRC, serve the function of extracting a suitable number of coded bits to match the time/frequency resources assigned for transmission as well as that of generating different redundancy versions needed for effective HARQ operation. Rate matching is performed separately for each code block and attached CRC and consists of bit selection and bit interleaving.

8.6.4.3 Scrambling

Each block of coded bits created by the HARQ functionality is scrambled. Scrambling is achieved, as shown in Fig. 8.10, by generating a repetitive but long pseudorandom bit sequence and logically combining the generated sequence with incoming data. The scrambled output assumes properties similar to the pseudorandom sequence, irrespective of input data properties. The purpose of scrambling here is intercell interference suppression. With the same spectrum being likely used in all neighboring cells, such interference is a real possibility. Without scrambling, the channel decoder could possibly be as equally matched to an interfering signal as to the desired signal using the same frequency resource and thus be unable to properly suppress the interference. Different scrambling sequences are applied in the DL to neighboring BSs and in the UL to different UEs. As a result, interfering signals, after descrambling, are randomized making them appear more noise like and thus more easily addressed via the channel decoding.

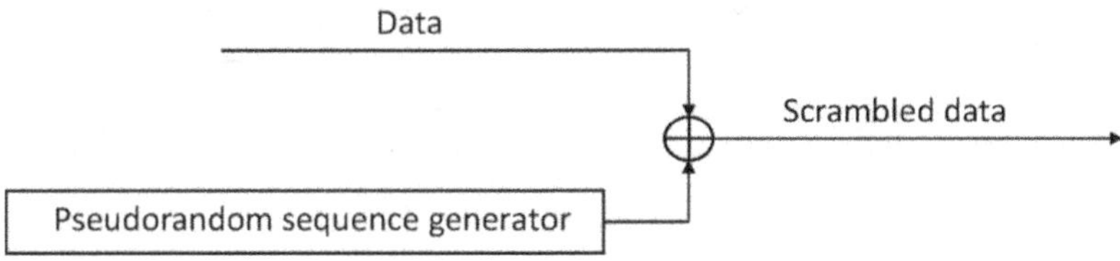

Fig. 8.10 Scrambler

8.6.4.4 Resource Element Mapping

Following scrambling, the block of scrambled bits undergoes modulation symbol mapping (Sect. 6.2.1) followed by layer mapping (Sect. 7.3.3). If it is UL processing using optional DFTS-OFDM (Sect. 6.4), then layers received from the layer mapper are DFT spread at this stage. If UL or DL processing is using CP-OFDM, then this stage is skipped. Layers then undergo multi-antenna precoding (Sect. 7.3.3) followed by the resource element mapping stage, the subject of this section.

A resource element is the smallest physical resource in an OFDM time-frequency resource grid. It represents one modulation symbol (time) by one subcarrier (frequency). In NR a *resource block* (RB) is one-dimensional in the frequency domain only and is 12 subcarriers wide. Resource mapping [2] is the taking of modulation symbols to be transmitted on each antenna port and mapping them to the set of resource elements available in a set of resource blocks dictated by the MAC scheduler. It is here that each subcarrier is modulated with a modulation symbol to create a modulated subcarrier.

8.6.4.5 Transmission Signal Duplexing

In Fig. 8.8, for purposes of simplicity, up-conversion is omitted. In real systems, however, baseband OFDM signals are up-converted to RF (Sect. 6.2.1), where prior to transmission they undergo antenna duplexing so that one antenna can serve both transmission and reception. In this section, we review antenna-duplexing schemes.

A transmission signal duplexing scheme is a method of accommodating the bidirectional communication of signals between two devices. In mobile access systems, this means the bidirectional communication between the BS and a UE. There are three duplexing schemes that are utilized in mobile access networks, namely, Frequency Division Duplexing, Frequency Switched Division Duplexing, and Time Division Duplexing.

Frequency Division Duplexing (FDD) is the traditional form of duplexing. In PMP systems utilizing FDD, the DL and UL frequency channels are separate, and thus all UEs can transmit and receive simultaneously. The channels bandwidths are usually, but not necessarily, of equal size. For proper operation of FDD is necessary that there be a large separation between the two assigned frequencies. This is

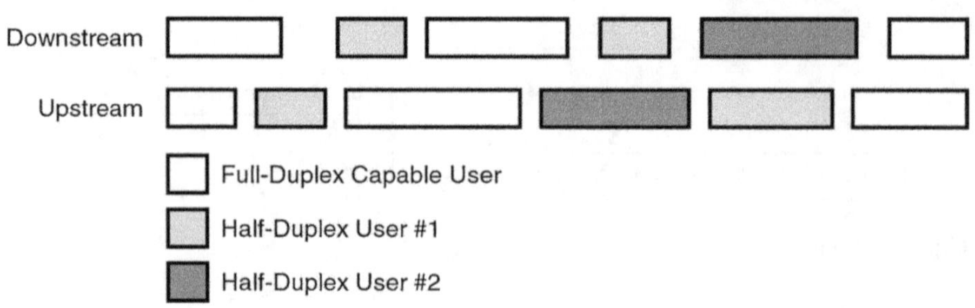

Fig. 8.11 Example full-duplex and half-duplex time allocation

because, at a given BS or UE, it is necessary that the receiver be able to filter out the transmit signal sufficiently to prevent it from interfering with the received signal.

Frequency Switched Division Duplexing (FSDD), also known as *half-duplex-Frequency Division Duplexing* (*H-FDD*), is a duplexing scheme in which, like FDD, the DL and UL frequency channels are separate but where some or all of the UEs cannot transmit and receive simultaneously but must do so sequentially. Those UEs that can operate simultaneously are said to operate in a *full-duplex* mode, whereas those that must operate sequentially are said to operate in a *half-duplex* mode. The design of UEs that operate in a half-duplex mode is simplified as the antenna-coupling device now consists essentially of only a switch that switches between the transmitter output and receiver input instead of a frequency-separating device. An example of time allocation on a FSDD system is shown in Fig. 8.11.

Time Division Duplexing (TDD) is a duplexing scheme where the downstream and upstream transmissions share the same frequency channel but do not transmit simultaneously. However, because of the rapid speed of switching between these transmissions, "simultaneous" two-way communication is preserved. Whereas FDD transmission requires a large frequency separation, TDD requires a guard interval between transmission and reception. This interval must be large enough to allow for, among other factors, a transmitted signal to arrive at its intended terminal before a transmission at that terminal is started, thus switching off reception. The guard interval is typically on the order of 100 to a few hundred microseconds, with the higher end values being required for cells covering large distances. Note, however, that large guard intervals reduce system capacity as the guard time is not available for useful transmission. The allocation of time between downstream and upstream traffic is normally adaptive, making this technique highly attractive for situations where the ratio of downstream to upstream traffic is likely to be asymmetric and highly variable. A significant advantage of TDD over FDD is channel reciprocity, i.e., both the DL and UL channels share the same propagation characteristics. As a result, the BS is able to estimate the DL channel from its measurement of the UL channel, removing the need to feed this information from the UE up to the BS. One impediment to the deployment of TDD systems is that many global spectrum allocations dictate FDD and prohibit TDD because of the difficulty in coordinating it with FDD systems from an interference point of view. This difficulty arises because the transmission of the same frequency in both DL and UL directions makes the discrimination of a nearby antenna to these signals limited to non-existent.

The duplexing scheme applied is dictated by the allocation of spectrum by the governing authority. Allocation is either paired, thus facilitating FDD, or unpaired, facilitating TDD. In general, bands below about 10 GHz tend to be, but not always, paired. In the millimeter wave range, on the other hand, assigned bands are increasingly unpaired. 5G NR can operate in both FDD and TDD mode. FDD will be the main duplexing scheme in the below 10-GHz bands and TDD more common in the millimeter wave bands. Because of their high signal loss, systems operating in the millimeter wave range will typically cover small cells with a relatively small number of users per BS. In such situations, the DL/UL traffic requirements can vary rapidly. To make optimum use of available total DL/UL capacity, NR uses *dynamic TDD* which allows the dynamic assignment of DL and UL resources, i.e., the ratio of time assigned to DL traffic to that assigned to UL traffic varies dynamically according to need.

8.6.4.6   Carrier Aggregation

Carrier aggregation (CA) [2, 5] refers to the concatenation of multiple carriers that can be transmitted in parallel to and from the same device thus increasing the available transmission bandwidth and in turn maximum data rate achievable. Aggregated carriers are called component carriers. As shown in Fig. 8.12, component carriers

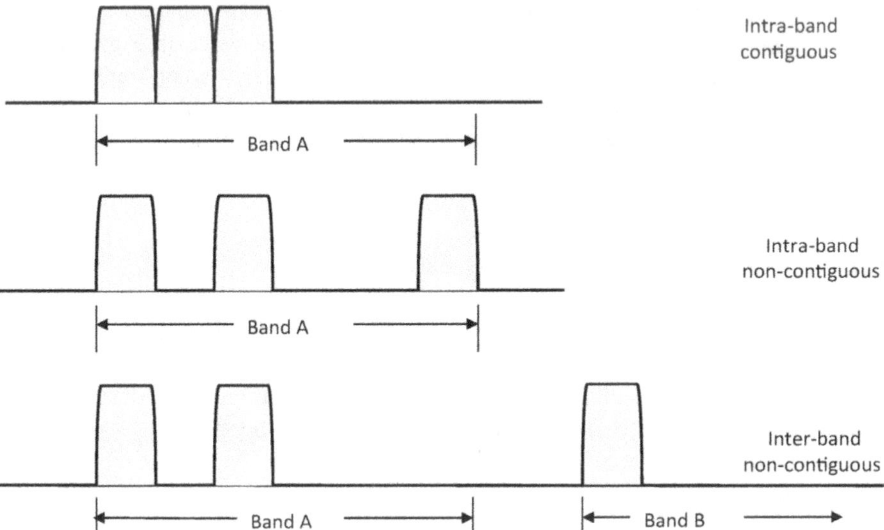

Fig. 8.12 Types of carrier aggregation

need not be contiguous and in the same frequency band but can be aggregated in one of three scenarios:

- Intra-band aggregation with contiguous component carriers.
- Intra-band aggregation with non-contiguous component carriers.
- Inter-band aggregation with non-contiguous component carriers.

5G NR supports carrier aggregation of up to 16 component carriers, in both the below 10-GHz range and the millimeter wave range, component carriers having the same or different numerologies, and possibly of different bandwidths and duplexing schemes. As the maximum NR channel bandwidth in the FR2-1 range (Sect. 8.8) is 400 MHz, the maximum aggregated bandwidth possible, but highly unlikely, in this range is 16×0.4 GHz $= 6.4$ GHz!

The number of component carriers in any aggregation is independently configured for DL and UL. In the inter-band aggregation scenario where there are multiple TDD carriers, the transmission direction of the different carriers need not be the same. This implies the possibility of a TDD device capable of carrier aggregation transmitting on one frequency while receiving on another and thus needing an RF duplexer.

In the NR specification, each component carrier appears as a separate "cell," and a carrier aggregation capable UE is said to be able to communicate with multiple cells. Such a UE connects to one primary serving cell (PCell) and one or more secondary serving cells (SCells). The PCell is the cell that the UE initially selects and is connected to. Once connected to the BS via the PCell, one or more SCells can be configured. Once configured, secondary cells can be rapidly activated or deactivated based on need.

A single scheduler entity in the BS MAC schedules all UEs and all their corresponding component carriers. At the physical layer, as depicted in Fig. 8.9, each transport block is mapped to a single component carrier. Even when multiple component carriers are scheduled on a UE simultaneously, coding, HARQ, modulation, numerology, and resource element mapping, along with corresponding signaling, are performed independently on each component carrier.

8.6.4.7 Pertinent Physical Layer Specifications

Starting at transmitter input, pertinent physical layer specifications are as follows:

CRC processor: Receives transport blocks from the MAC layer. For transport blocks larger than 3824 bits, a 24-bit CRC is added. For transport blocks less than or equal to 3824 bits, a 16-bit CRC is used to reduce overhead.

Coding: LDPC coder. See Sect. 5.3.4.

Modulation symbol mapper:

- For FR1 DL with CP-OFDM: QPSK, 16-QAM, 64-QAM, 256-QAM, 1024-QAM.
- For FR2 DL with CP-OFDM: QPSK, 16-QAM, 64-QAM, 256-QAM.

- For FR1 UL with CP-OFDM: QPSK, 16-QAM, 64-QAM, 256-QAM.
- For FR1 UL with DFTS-OFDM: π/2 BPSK, QPSK, 16-QAM, 64-QAM, 256-QAM.
- For FR2 UL with CP-OFDM: QPSK, 16-QAM, 64-QAM.
- For FR2 UL with DFTS-OFDM: π/2 BPSK, QPSK, 16-QAM, 64-QAM.
  Note 1: DFTS-OFDM is optional and used in link budget-limited cases.
  Note 2: Maximum modulation orders shown as per Release 17, where such maximum orders here means that 3GPP has specified for those orders "reference measurement channels" used to carry out specified test requirements.

Layer mapper:

- For the DL: One coded transport block mapped on up to four layers. If there is a second coded transport block, it is mapped on up to four more layers.
- For the UL: One coded transport block mapped on up to four layers.

OFDM signal generator:

- FFT size: 512 minimum, 4094 maximum.
- Subcarrier spacing: 15, 30, 60,120, 480, and 960 kHz.

Resource and physical antenna mapper: Broadly speaking, modulation symbols to be transmitted on each antenna port can be mapped to any resource element not specifically designated to other physical channels or signals. Resource elements in turn are mapped to the 1000 series antenna port(s) on the DL and 0 series antenna ports on the UL.

In the receiver, the processing is the inverse of that in the transmitter.

8.7   5G Data Rates

Extremely high data rates are one of the major features of 5G NR. For NR, the approximate maximum data rate per component carrier can be computed as follows [6]:

$$\text{Maximum data rate}\left(Mb/s\right) = \frac{1}{T_s^{\mu}} \times N_{PRB}^{BW,\mu} \times$$
$$12 \times \left(1 - OH\right) \times Q \times R_{max} \times f \times v \times 10^{-6} \tag{8.1}$$

where:

- $T_s^{\mu} = {10^{-3}}/{14 \times 2^{\mu}}$ is the average OFDM symbol duration in a subframe of numerology μ, assuming normal cyclic prefix.
- $N_{PRB}^{BW,\mu}$ is the maximum RB allocation in the available system bandwidth BW with numerology μ, as defined in Sect. 5.3.2 of [6] and as given in Table 8.2 for minimum and maximum bandwidths.

– OH is the time–frequency resource overhead. Simply stated, this is the average ratio of all the resource elements not used by the PDSCH or the PUSCH to the total number of available resource elements. It takes the following values:

0.14 for FR1 DL
0.18 for FR2 DL
0.08 for FR1 UL
0.10 for FR2 UL

– Q is the bits per symbol for the applied modulation scheme, being 10 for 1024-QAM, the highest order modulation supported.
– $R_{\max}$ is the maximum code rate. In 5G NR, it is 948/1024.
– f is a scaling factor used to reflect the capability mismatch between baseband and RF capability of the UE. It is signaled per band and can take the values 1, 0.8, 0.75, and 0.4.
– v is the maximum of layers.

We now consider why the above equation gives maximum data rate.

– T_s^{μ} : The subframe (slot) duration, as indicated in Sect. 8.6.2, is given by $\dfrac{10^{-3}}{2^{\mu}}$ s. Thus, assuming 14 OFDM symbols per slot, the OFDM symbol duration is given by $\dfrac{10^{-3}}{14 \times 2^{\mu}}$, and $1/T_s^{\mu}$ represents the *OFDM symbol* rate.
– One OFDM symbol is composed of $N_{PRB}^{BW,\mu} \times 12$ modulated subcarriers. Thus, $\dfrac{1}{T_s^{\mu}} \times N_{\mathrm{PRB}}^{\mathrm{BW},\mu} \times 12$ represents the total *modulating symbol* rate, $\mathrm{SR_T}$ say.
– $(1 - \mathrm{OH})$ represents the fraction of Res available for user data transmission. Thus, $\mathrm{SR_T} \times (1 - \mathrm{OH})$ represents the fraction of modulating symbols available for user data.
– The user bits per OFDM modulating symbol equals $Q \times R_{\max}$. Thus, the user *bit* rate is $\mathrm{SR_T} \times (1 - \mathrm{OH}) \times Q \times R_{\max}$.
– v: Data rate clearly proportional to the number of layers hence the multiplication by v.
– f: This scaling factor allows the introduction of a practical baseband to RF limitation to the calculation.

Release 15 specified 256-QAM as the modulation with the highest order for FR1 in the DL and UL, and 64-QAM as the modulation with the highest order for FR2 in the DL and UL. Release 16 changed the specified modulation with the highest order in the FR2 DL to 256-QAM. Release 17 changed the specified modulation with the highest order in the FR1 DL to 1024-QAM. Note that "specified" here means that 3GPP has specified a "reference measurement channel" used to carry out specified test requirements. Table 8.3 shows the maximum NR data rates per layer ($v = 1$), per component carrier, data being computed via Eq. (8.1) and assuming FR1 DL modulation of 1024-QAM, FR2 DL modulation of 256-QAM, FR1 UL modulation of 256-QAM, FR2 UL modulation of 64-QAM, and a scaling factor of 1.

Table 8.3 Maximum NR data rates per layer per component carrier

Frequency range	Subcarrier spacing (kHz)	Bandwidth (MHz)	Downlink rate (Mb/s)	Uplink rate (Mb/s)
FR1	15	50	361	309
FR1	30	100	730	625
FR1	60	100	722	618
FR2-1	60	200	1078	887
FR2-1	120	400	2155	1774
FR2-2	480	1600	8097	6665
FR2-2	960	2000	9664	7955

For downlink transmission, the maximum number of layers is 8, thus the DL rates shown in Table 8.3 must be multiplied by 8 to get the maximum achievable DL rate per component carrier. For the UL, to maximum number of layers is 4, thus the UL rates shown in Table 8.3 must be multiplied by 4 to get the maximum UL achievable rate per component carrier. Via carrier aggregation (Sect. 8.6.4.6), data can be carried on multiple component carriers simultaneously. Thus, the maximum data rate possible is the sum of the maximum data rates of the individual component carriers that are aggregated.

All FR2 bands are TDD. Thus, with operation in such bands, the maximum downlink and uplink rates are not mutually exclusive. In typical scenarios, there is more downlink demand than uplink demand. Such a scenario could be, for each slot, ten ODFM symbols for the downlink, one symbol for transition, and three symbols for the uplink. This would result in peak downlink rates of 71% of the maximum available and peak uplink rates of 21% of maximum available.

It is important to not lose sight of the fact that the data rates given in Table 8.3 represent the theoretically best achievable with NR specifications and that in the real world will only likely be achieved by a small percentage of UEs. These rates assume that in the DL and UL modulation is the maximum supported order and the coding rate is the maximum supported of 984/1024. This will only be the case for high SINR situations and thus likely only for UEs located close to the base station. As UE distance from the base station increases, both the modulation order and coding rate will decrease to the point where at the cell edge the modulation is likely to be QPSK and the coding rate less than maximum. Even if the coding rate stays at the maximum, the peak rate attainable with QPSK would only be one fifth that attainable with 1024-QAM. As a result, very roughly speaking, the average maximum data rate per layer per component carrier attainable by a UE in both the DL and UL will likely be less 50% of the maximum rates shown in Table 8.3.

Clearly, in a multiuser system, where total capacity has to be shared by all users, practical user average data rates will be significantly less than the maximum rates discussed above. In such situations, the average user data rate decreases as the number of users increases but is also dependent on bandwidth, received signal levels, and intercell interference levels.

8.8   Spectrum for 5G

5G NR is conceived to operate ultimately in spectrum ranging from approximately 400 MHz up to about 90 GHz. It is being designed to be able to operate in licensed, unlicensed, and shared frequency bands, shared bands being those where 5G shares the spectrum with nonmobile service providers.

3GPP originally defined two frequency ranges for NR, namely, FR1 and FR2. FR1 was specified in 3GPP TS 38.104 V15.1.0 to cover the range 410–6000 MHz. However, in V15.5.0, the range was extended covering 410–7125 MHz to allow it to be used in unlicensed bands in the 6-GHz region. FR2 was specified in 3GPP 38.104 V15.1.0 to cover the range 24.25–52.6 GHz, frequency bands in this range being referred to as millimeter wave (mmWave) bands. Spectrum wise, the FR2 range is the distinguishing feature of 5G relative to 4G. It allows for transmission in bands with much more spectrum than available in FR1 enabling high capacity and data rates.

In 3GPP TS 38.104 V17.6.0, the FR2 range was extended and split in two subranges, FR2-1 covering the original FR2 range of 24.25–52.6 GHz and FR2-2 covering the new subrange of 52.6–71 GHz.

New bands are continuously being defined by 3GPP for NR. Band assignment is not only for operation in the FDD or TDD mode (Sect. 8.6.4.5), but may be for operation in the *supplementary uplink* (SUL) or *supplementary downlink* (SDL) mode.

SUL is used to extend the UL coverage. With SUL operation there is a standard FDD band assignment, with DL and UL bands as well as a supplementary uplink band, the SUL band being of a lower frequency than the standard UL band. By being of a lower frequency path loss is lowered, resulting in greater coverage. When UL coverage is acceptable on the standard band, UL transmission is on this band. If UL coverage starts to deteriorate, the UE dynamically selects the SUL band for transmission instead of the standard UL band. The SUL band cannot be used alone but only in conjunction with a standard FDD band. SDL is used to increase DL capacity and thus provide additional data transmission capability to UEs. With SDL operation there is a standard FDD band assignment, with UL and DL bands as well as a SDL band which is carrier aggregated (Sect. 8.6.4.6) with the standard DL band.

8.8.1   Licensed FR1 and FR2-1 5G Spectrum

In Release 15 only operation in licensed bands is supported. The lowest licensed frequency bands currently specified in FR1 is for FDD operation, covering approximately 600–800 MHz. Total spectrum in these bands is on the order of 15–40 MHz per direction and thus, as we shall see below, result in much lower capacity relative to

FR2 bands. The big advantage of such bands, however, is wide area coverage, typically on the order of tens of kilometers, and low outdoor-to-indoor penetration loss.

In the licensed 1.4–2.7 GHz region of FR1, the operation is mostly FDD with typical maximum spectrum per operating frequency being 20 MHz per direction. With carrier aggregation, up to 16 such carriers can theoretically be aggregated. Such a large number of aggregated carriers is unlikely, however, with a typical maximum more likely to be about 5 for a total aggregated bandwidth of about 100 MHz. Bands available in this range have been widely used by 3G and 4G networks.

In the licensed 3.3–4.2 GHz region of FR1, often referred to as the C-Band or the 3.5-GHz band, the operation is TDD, and the maximum spectrum per operator is typically about 100 MHz. This band is likely to be highly used in 5G networks as the amount of spectrum is relatively large yet propagation loss only on the order of about 5 dB more than those bands in the 2-GHz range. Furthermore, this additional path loss can be overcome by utilizing high-gain beamforming antennas leading to coverage similar to that in the 2-GHz bands.

In the licensed FR2 range, all operation is TDD, and available spectrum undergoes a large increase relative to FR1. In the bands in this range, the spectrum per operator range typically up to 400 MHz which supports user rates of up to 5 Gbps. The price paid for these high data rates is much reduced coverage, typically on the order of hundreds of meters or less. Furthermore, this low-coverage problem is made more difficult due to achievable base station output power being generally lower than that in FR1 bands and penetration loss through walls, windows, and doors being much higher than in the lower bands, as discussed in Sect. 3.2.4.6.

Table 8.4 provides licensed operating bands in the 600–800-MHz portion of FR1. It will be seen that the maximum per carrier bandwidth is 35 MHz. Table 8.5 provides those NR licensed operating bands in the mid to high portions of FR1 that afford the largest channel bandwidth, i.e., 100 MHz. These bands are therefore likely to be very desirable. Finally, Table 8.6 presents NR licensed operating bands in FR2 where the maximum per carrier bandwidth is 400 MHz.

Table 8.4 NR licensed operating bands in 600–800 MHz portion of FR1 as of Release 17, V17.6.0

NR band #	Uplink frequency range (MHz)	Downlink frequency range (MHz)	Duplex mode	Maximum channel bandwidth (MHz)	Release
n12	699–716	729–746	FDD	15	15
n13	777–787	746–756	FDD	10	17
n14	788–798	758–768	FDD	10	16
n28	703–748	758–803	FDD	20	15
n29	N/A	717–728	SDL	10	16
n67	N/A	738–758	SDL	20	17
n71	663–698	617–652	FDD	35	15
n83	703–748	N/A	SUL	30	17
n85	698–716	728–746	FDD	15	17

Table 8.5 Selected licensed NR operating bands in mid to high portions of FR1

NR band #	Uplink frequency range (MHz)	Downlink frequency range (MHz)	Duplex mode	Maximum channel bandwidth (MHz)	Release
n40	2300–2400	2300–2400	TDD	100	15
n41	2496–2690	2496–2690	TDD	100	15
n48	5150–5925	5150–5925	TDD	100	16
n77	3300–4200	3300–4200	TDD	100	16
n78	2200–3800	3300–3800	TDD	100	15
n79	4400–5000	4400–5000	TDD	100	15
n97	2300–2400	N/S	SUL	100	17
n104	6425–7125	6425–7125	TDD	100	17

Table 8.6 NR licensed operating bands in FR2-1 as of Release 17, V17.6.0

NR band #	Uplink frequency range (GHz)	Downlink frequency range (MHz)	Duplex mode	Maximum channel bandwidth (MHz)	Release
N257	26.50–29.50	26.50–29.50	TDD	400	15
N258	24.25–27.50	24.25–27.50	TDD	400	15
N259	39.50–43.50	39.50–43.50	TDD	400	16
N260	37.00–40.00	37.00–40.00	TDD	400	15
N261	27.50–28.35	27.50–28.35	TDD	400	15
N262	47.20–48.20	47.20–48.20	TDD	400	17

8.8.2   Unlicensed FR1 5G Spectrum

In Release 16 support was expanded to include operation in unlicensed spectrum, targeting the 5-GHz and 6-GHz bands. Support is for *licensed-assisted access* (LAA) as well as standalone unlicensed operation. With LAA, the mobile device is connected to the network via a licensed carrier which handles initial access and mobility as well as one or more unlicensed carriers used to increase the capacity and data rate. With standalone unlicensed operation, there is no connection via a licensed carrier and thus initial access and mobility are handled entirely via unlicensed operation. Because unlicensed bands are available to all they are accessed on a first come first serve basis, where a potential user must first listen to check if the band is free and only access it if it is. Thus, though the large spectrum potentially affords high data rates, operation is subject to high latency which can lower the average achieved data rate. Furthermore, operation in these bands is subject to rules, including the maximum transmission power permitted. As this maximum power is usually less than that permitted in licensed bands, unlicensed operation is less suitable for wide area communication.

The 5-GHz band is available almost worldwide and is divided into two parts. The lower part, 5150–5350 MHz, is typically used indoors as the maximum transmission power permitted here is usually 23 dBm. The upper part, 5470–5925 MHz (upper limit a bit higher in some regions), is typically used indoor and outdoor, transmission power of up to 30 dBm normally being allowed. Over and above maximum power limitations, additional rules are imposed in some parts of this band in some regions, these rules including:

- Listen-before-talk, a requirement that the transmitter listens for any activity on the channel before each transmission so as to not commence transmission when the channel is occupied by another user.
- Power spectral density limitations, resulting in the device not being allowed to transmit at full power when operating in a bandwidth less than a specified value.
- Maximum channel occupancy time, this limiting to period of continuous transmission to a stated amount.
- Dynamic frequency selection, used to vacate the channel upon the detection of interference from radar systems.
- Transmitter power control, a requirement that the transmitter be capable of lowering its output power, typically by 3 or 6 dB relative to its maximum permitted, in order to reduce the overall interference level as required.

The 5-GHz band, 5150–5925, is available almost worldwide and has been designated by 3GPP for unlicensed NR operation in the TDD duplex mode. The entire band is listed as channel n46 and is restricted to LAA operation.

The 6-GHz band, 5925–7125 MHz, has been listed by 3GPP for NR operation in the TDD duplex mode. The entire band is listed as channel n96 with LAA operation in the United States and certain other countries, and the partial band 5925–6425 MHz listed as channel n102, also with LAA operation, in Europe and certain other countries. These bands are currently applicable only in the U.S. and other countries that allow LAA. In the United States, the specifications on the unlicensed use are designed to minimize interference with licensed use. Specifically, two different types of unlicensed operation are authorized, indoor low-power and outdoor standard power. The indoor units have a maximum allowed Equivalent Isotropic Radiated Power (EIRP) (Sect. 3.2.1) of 30 dBm (1 W) and the standard-power units are allowed a maximum EIRP of 36 dBm (4 W). Low-power operation is permitted all across the band and its power level is such that it is unlikely to cause interference with licensed outdoor links. For those parts of the band that support a large number of high reliability point-to-point links such as mobile backhaul links standard-power operation is permitted under the control of an *automated frequency coordination* (AFC) system. The AFC system determines the frequencies on which standard-power devices can operate without likely causing harmful interference to incumbent microwave receivers. It does this by referring to a database of incumbent licensed systems and then makes only available for use by standard-power devices those frequencies that won't interfere with incumbent links. However, this cannot fully guarantee that outages due to interference will not occur. In Europe, the n102 was opened in 2021 to unlicensed operation but only at power levels suitable for Wi-Fi.

Table 8.7 presents the unlicensed bands designated in FR-1.

Table 8.7 NR unlicensed operating bands in FR-1

NR band #	DL/UL band	Duplex mode	Release
n46	5150–5925	TDD (Note 1)	16
n96	5925–7125	TDD (Note 1)	16
n102	5925–6425	TDD (Note 1)	17

Note 1: Restricted to shared spectrum access (LAA operation)

8.8.3 Unlicensed FR2-2 5G Spectrum

3GPP in Release 17 added to its list of approved bands the band 57.0–71.0 GHz (n263) for unlicensed NR operation in the TDD mode. This band falls within the range of FR2-2, offers wide transmission bandwidths ranging from 100 to 2000 MHz and hence the possibility of very high transmission data rates. However, it is not without issues. For example, in the 60-GHz region, there is very large oxygen absorption which severely limits transmission distance.

8.9 Transmitter Output Power and Receiver Reference Sensitivity

There are hundreds of individual specifications that define the parameters of 5G base stations and UEs, all precisely stated in the 3GPP 5G NR technical specifications. Two performance parameters, however, have a large impact on coverage, these being transmitter output power and receiver sensitivity. The difference between the two, plus transmitter and receiver antenna gain, defines the maximum path loss tolerable while maintaining packet error rate above the defined minimum. In an ideal world, one thus seeks, within limits, to maximize transmitter output power and minimize the value of receiver sensitivity. This section will review at a high-level NR specified base station and UE transmitter output power and receiver reference sensitivity, the latter being a specified receiver sensitivity in a defined reference channel.

8.9.1 Base Station Transmitter Output Power

3GPP base station specifications apply to three classes, namely Wide Area, Medium Range, and Local Area, and to four station types, namely BS type 1-C, BS type 1-H, BS type 1-0, and BS type 2-0 [7]. Type 1-C, 1-H, and 1-0 apply to FR1 while type 2-0 applies to FR2.

For type 1-C base stations, "conducted" requirements are applied at the antenna connector. Should external equipment such as a power amplifier and or a filter be used, then requirements are applied at the far end antenna connector.

Table 8.8 Base station conducted output power

BS class	BS type 1-C	BS type 1-H	
	$P_{rated,c,AC}$	$P_{rated,\,c,TABC}$	$P_{rated,c,sys}$
Wide area BS	No upper limit	No upper limit	No upper limit
Medium range BS	$\leq$38 dBm	$\leq$38 dBm	$\leq$38 + 10 log N dBm
Local area BS	$\leq$24 dBm	$\leq$24 dBm	$\leq$24 + 10 log n dBm

Note: (1) $P_{rated,c,AC}$: The rated total output power declared at the antenna connector. (2) $P_{rated,c,TABC}$: The rated carrier output power per TAB connector. (3) $P_{rated,c,sys}$: The sum of $P_{rated,c,TABC}$ for all TAB connectors for a single carrier. (4) N is the number of active transmitter units

For type 1-H base stations the transceiver unit array (one or more transceivers) is connected to a composite antenna that includes an antenna array. The requirements are specified at two reference points, signified by *conducted requirements* and *radiated requirements*. Conducted characteristics are specified at the connector/connectors at the transceiver array boundary (TAB). Radiated characteristics are specified over the air (OTA).

For types 1-0 and 2-0 characteristics are specified only OTA. For type 1-0 base stations, the transceiver unit array must contain at least 8 transmitter units and at least 8 receiver units. Type 2-0 is defined for FR2.

Base station *conducted* transmitter output power is as specified in Table 8.8.

For *radiated* transmitter output power:

- Radiated transmitter power is defined as the equivalent isotropic radiated power (EIRP) level for a declared beam at a specified beam peak direction. OTA BS output power is declared as the total radiated power (TRP) requirement.
- For type 1-0, for the Wide Area class, there is no upper limit on the rated carrier TRP.
- For type 1-0, for the Medium Range and Local Area classes, the rated carrier TRP is $\leq$47 and $\leq$33 dBm, respectively.
- For type 2-0, there is no upper limit for the rated carrier TRP.

8.9.2 Base Station Receiver Reference Sensitivity

The conducted reference sensitivity is the minimum mean power received at the antenna connector for base station type 1-C and at the transceiver array boundary for type 1-H at which a throughput requirement shall be met for a specified reference measurement channel [7], this throughput being $\geq$95% of the specified maximum throughput of the reference measurement channel. Many conducted reference sensitivities are specified depending on base station class and operating band. To get a sense of their values we note that for the ensemble of base stations operating in licensed bands, reference conducted sensitivity varies from $-$87.3 to $-$101.7 dBm. For the ensemble of base stations operating in unlicensed bands, conducted reference sensitivity varies from $-$86.3 to $-$103 dBm.

OTA reference sensitivity power level for BS type 1-0 and 2-0 is the minimum mean power received at the "radiated interface boundary" at which a reference performance requirement shall be met for a specified reference measurement channel. The reference performance requirement is that the throughput shall be $\geq$95% of the maximum throughput of the reference measurement channel as specified. The specified levels [7] are many and varied, are a function of BS channel bandwidth, subcarrier spacing, and the reference measurement channel and will not be stated here.

8.9.3 UE Transmitter Output Power

For FR1, UE maximum output is specified per power class [8], there being four such classes:

- Power Class 1: power 31 dBm, band n14.
- Power Class 1.5: power 29 dBm, bands n41, n77, n78, n79.
- Power Class 2: power 26 dBm, bands n34, n39, n40, n41, n77, n78, n79.
- Power Class 3: power 23 dBm, default power class and covers all bands.

  For FR2, seven power classes are specified [9]:

- Power Class 1, Fixed wireless access (FWA): power 35 dBm for all bands except n259, n263.
- Power Class 2, Vehicular UE: power 23 dBm for all bands except n260.
- Power Class 3, Handheld UE: power 23 dBm for all bands except n263, 27 dBm for band n263.
- Power Class 4, High-power non-handheld UE: power 23 dBm for all bands except n259, n263.
- Power Class 5, Fixed wireless access (FWA): power 23 dBm for bands n257, n258, n259.
- Power Class 6, High Speed Train Roof-Mounted: power 23 dBm for bands n257, n258, n261.
- Power Class 7, RedCap (See Sect. 8.12.2): power 23 dBm for mmWave bands n257, n258, n261.

  For all power classes, a maximum EIRP is specified as 20 dB above the maximum output power. This implies a maximum antenna gain of 20 dB.

8.9.4 UE Receiver Reference Sensitivity

In FR1, the UE is required to be equipped with a minimum of two receive antenna ports in certain operating bands and with a minimum of four receive antenna ports in others. Here, the reference sensitivity power level is the minimum mean power applied to each one of the UE antenna ports at which there is a sufficient SINR for the specified reference measurement channel to achieve a throughput that's $\geq 95\%$ of the maximum possible [8]. To give a sense of the sensitivity power levels, for two antenna port UEs with FDD band operation and QPSK modulation, the lowest level indicated is -100 dBm and occurs in the lowest bandwidth channel (5 MHz) in operating bands n1, n24, n70, and n100. The highest level indicated is -78.4 dBm and occurs in the 35-MHz bandwidth channel in operating band n8.

In FR2, as with FR1, reference sensitivity is specified for a specified reference measurement channel, with levels given are per power class for the various operating bands and channel bandwidths [9]. The lowest level indicated is -97.5 dBm in Power Class 1 (FWA) for a 50 MHz bandwidth channel in bands n257, 258, and n261. The highest level indicated is -73.8 dBm in Power Class 3 (handheld UE) for a 400-MHz bandwidth channel in band n262.

8.10 5G NR Multiple Antenna Options

Multi-antenna transmission options specified for 5G NR [2] build on, and in some cases replace, those options specified for LTE networks. The main driver for advanced techniques is the inclusion of mmWave communications in 5G systems which brings with it higher propagation losses and increased thermal noise power in cases of wider bandwidths. The key addition to address this change is steerable beamforming via antenna arrays with a very large number of elements which provide high gain and hence allows extended coverage. At the sub-6-GHz bands, the main improvement is mMIMO coupled with more focused beamforming. This combination, via improved spatial separation and thus minimized intercell interference, improves cell spectral efficiency by allowing an increase in the number of simultaneous full resource users. Note, however, that at these lower frequencies, the physical size may limit the number of elements in an antenna array since the area occupied by each element is proportional to the square of the wavelength.

MIMO transmission and beamforming of both control and traffic channels is a distinguishing feature of 5G NR. In LTE, the DL control channels use transmit diversity to ensure sufficient link budget, while the traffic channels use spatial multiplexing MIMO. With NR, the control channels rely on beamforming to achieve coverage. In sub 6-GHz bands, transmit diversity would probably have proved adequate in NR as it has in LTE. However, at mmWave bands, beamforming transmission is necessary to achieve reliable coverage, thus the pragmatic decision to use beamformed control channels throughout. Note, however, that NR supports specification transparent transmit diversity schemes in the UL.

For the DL, codebook-based closed loop precoding (Sect. 7.3.3) is specified. For SU-MIMO, NR in theory supports a maximum of eight layers. However, as handheld UEs only support four receive antennas, only a maximum of four layers per single handheld UE is possible. For MU-MIMO, a maximum of 16 layers in total is supported. As per Release 16, up to 4 layers per user is also possible. Thus, for example, one could have four UEs, each with four layers, or eight UEs, each with two layers.

For the UL, two modes of transmission are specified, namely, codebook based, and non-codebook based. In codebook-based transmission, the BS provides the UE with a transmit precoder matrix indication (PMI). The UE uses this indication to select the main UL data channel precoder from the codebook. In non-codebook-based transmission, which requires beam reciprocity, the precoder is determined locally based on received DL reference signals. Up through Release 17, up to four layers of SU-MIMO were supported when transmission is via OFDM-CP waveform. In Release 18, to better support non-smartphone devices such as vehicles and fixed wireless access customer premise equipment, this support is increased to eight layers via the support of eight transmit antennas. When transmission is via DFTs-OFDM waveform, only single-layer transmission is supported.

8.11   5G NR Release 16

Release 16, which took place in July 2020, added enhancements and new capabilities to Release 15. Among the new capabilities are:

- *NR access to unlicensed spectrum in the FR1 band* as already mentioned in Sect. 8.8.1.
- *Adding 256-QAM as a downlink option in the FR2 band.* In Release 15 FR2 DL modulation order was limited to 64-QAM. This addition increases the maximum DL data rate achievable by 33%.
- *Integrated Access and Backhaul* (IAB), an architecture that uses NR to provide wireless links between a central location and distributed cell sites and between cell sites (See Sect. 8.11.1).
- *Sidelink communication*, introduced as part of vehicle-to-anything (V2X) operation, but focused on vehicle-to-vehicle use case (See Sect. 8.11.2).
- *Multi-TRP transmission*, where DL data transmission via the PDSCH is transmitted simultaneously by a serving cell from two geographically separate locations (See Sect. 8.11.3).
- *Device positioning* based on Radio Access Technology (RAT) versus RAT independent techniques provided in Release 15 (See Sect. 8.11.4).

A comprehensive summary of Release 16 features can be found in [10, 11]. Following we review at a high-level IAB, sidelink communication, multi-TRP transmission, and NR positioning.

8.11.1   *Integrated Access and Backhaul (IAB)*

A key innovation in the evolution of 5G mobile networks is *Integrated Access and Backhaul* (IAB) [2, 12], a technology that is now standardized in 3GPP 5G NR Release 16. It permits the existing 5G access radio air interface technology to be the employed in the backhaul network, utilizing access spectrum for backhaul as well.

IAB efficiently integrates access and backhaul in the time, frequency and/or space domain and supports multi-hop backhauling. It can be in-band, where access and backhaul fully or partially share resources in the frequency domain, or it can be out-of-band where access and backhaul do not overlap in the frequency domain. In in-band operation, it can permit a more flexible use of spectrum via the dynamic sharing of it between access and backhaul as compared to a static allocation to both. However, this in-band sharing could impact network quality as a result of interference or could result in the reduction in access capacity.

Though IAB can be supported in both the FR1 (sub 7 GHz) and FR2 mmWave 5G NR access bands, the large amount of spectrum available in FR2 mmWave bands make them particularly attractive for IAB. Wireless operation in these bands is very supportive of small cell, dense deployments, where there is no existing fiber infrastructure. Deploying new fiber at the necessary density is costly, not only financially, but also in terms of time, as permitting and ensuing trenching can be a very lengthy process. Wireless deployment, on the other hand, can be cost effective and rapid.

Figure 8.13 shows typical 5G NR standalone (not also connected to the 4G core) IAB architecture. Key components of this architecture are the *IAB-donor* and the *IAB-node*. The IAB-donor is a gNB consisting of a CU and one or more DUs as described in Sect. 10.3 above, the donor in the figure showing two DUs. The donor CU connects to the core network via a NG link and can provide access to UEs directly as well as via IAB nodes. IAB-donor connections to UEs and IAB nodes are via the standard 5G NR over the air interface, i.e., the Uu interface. The links between the IAB donor and IAB nodes and between individual IAB nodes are

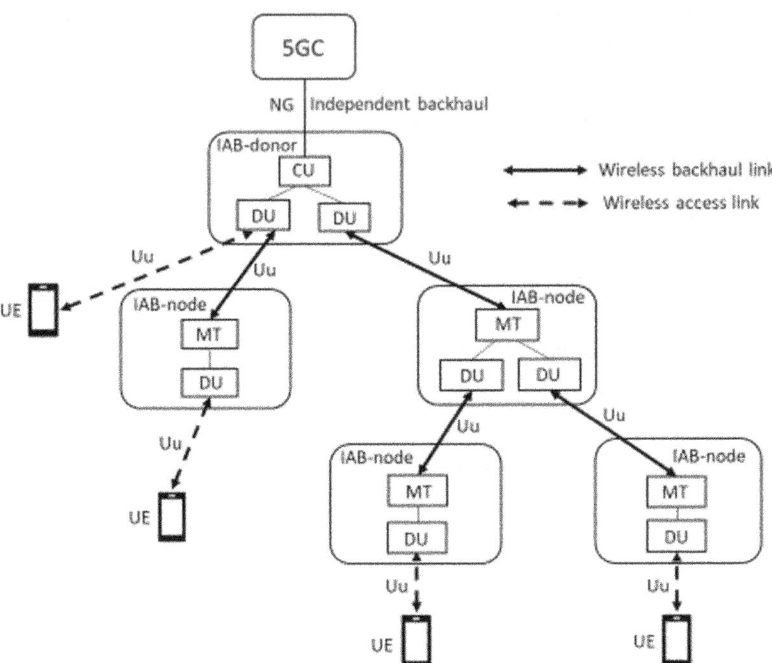

Fig. 8.13 Typical 5G NR stand alone IAB architecture

wireless backhaul links whereas those between UEs and either the IAB-donor or IAB nodes are wireless access links. The IAB-node consists of a *Mobile Termination* (MT) that's interconnected to a DU, the DU being as described in Sect. 8.3. The MT is a new entity whose function is to terminate the radio interface layers of the backhaul Uu interface toward the IAB-donor or another IAB-node. Both the IAB-donor and the IAB-node may contain multiple DUs. We note that the NG link between the IAB-donor and the 5GC is not provided as a part of IAB but rather is provided via an independent fiber or wireless backhaul network.

An important feature of IAB is that it supports multi-hop backhauling where an IAB-node can be backhauled to the IAB-donor via one or more IAB nodes. This is demonstrated in Fig. 8.13 where we see, for example, that the two lower IAB nodes are each backhauled to the uppermost IAB-donor via the IAB-node directly above. Note, however, that for Release 16, IAB-node mobility is limited to intra-CU mobility, i.e., an IAB-node can move between DUs of the same CU but not move into a cell covered by a different CU. IAB as defined in Release 16 is transparent to the UE.

8.11.2 Sidelink Communication

Sidelink (SL) communication [2, 13] refers to the direct communication between devices and was first introduced by 3GPP for LTE. It was not supported in the first NR release, Release 15, but was incorporated into Release 16 as part of the support for *Vehicle-to-Anything* (V2X) services, i.e., vehicle to vehicle, to pedestrian, to infra-structure units, to the network, etc. However, emphasis is on vehicle-to-vehicle services, these services falling into the following groups:

- *Vehicle platooning*: A group of vehicles traveling together organized into a platoon, with a lead vehicle communicating with the others allowing smaller intervehicle distances.
- *Extended sensors*: Exchange of sensor data between vehicles, pedestrians, etc., to extend the mobile unit's perception of the surrounding environment.
- *Advanced driving*: Autonomous or semi-autonomous driving via the exchange of sensor data and driving intention between vehicles enabling them to coordinate their trajectories.
- *Remote driving*: The remote driving of a vehicle for passengers who cannot drive themselves or a vehicle driven in a dangerous environment.

In Release 16, NR sidelink communication is, as indicated above, targeted toward vehicle-to-vehicle communication though this does not preclude its use for other cases. As shown in Fig. 8.14, three basic transmission scenarios are supported:

- *Unicast*: Here sidelink communication is with a specific device.
- *Groupcast*: Here sidelink communication is with a specific group of devices.
- *Broadcast*: Here sidelink communication is with any device that is within the range of transmission.

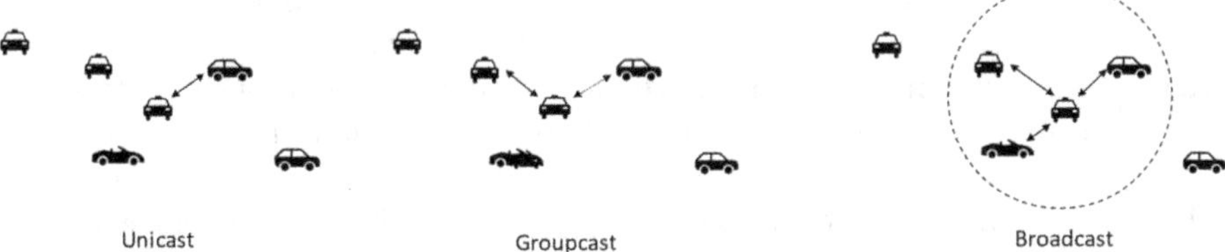

Fig. 8.14 Sidelink transmission scenarios

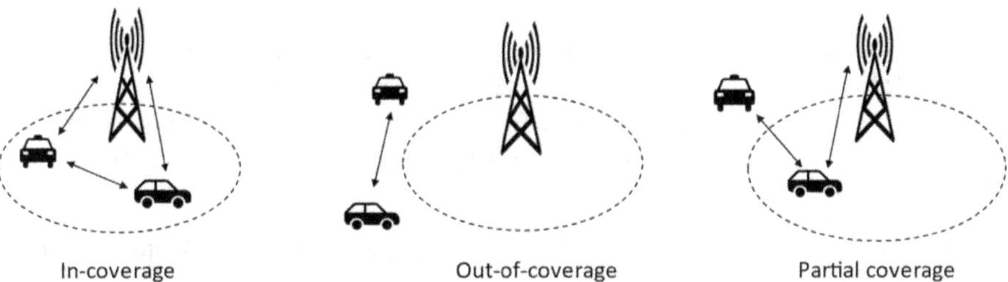

Fig. 8.15 Unicast transmission sidelink deployment scenarios

Regarding the relationship between the surrounding cellular network and sidelink communication, there are three deployment scenarios:

- *In-coverage operation*: Here the devices involved in sidelink communication are within the coverage of an overlaid cellular network.
- *Out-of-coverage operation*: Here the devices involved in sidelink communication are not within the coverage of an overlaid cellular network.
- *Partial coverage*: Here only one of the devices involved in sidelink communication is within the coverage of an overlaid cellular network.

Figure 8.15 shows these deployment scenarios for unicast transmission.

V2X messages can be transferred over the Uu interface, i.e., the interface between the gNB and the UE, or directly between UEs on a new interface, a sidelink interface known as PC5.

A device structured for sidelink transmission is configured with a resource pool [2], which specifies the overall time/frequency resource to be used for communication over a carrier. There are two basic modes of resource allocation:

- *Resource-allocation mode 1*: here an overlaid network schedules the transmission and is thus applicable only for in-coverage or partial coverage deployment scenarios. Scheduling is similar to how the network schedules for conventional uplink PUSCH transmission except that the scheduling grant is for sidelink, not uplink transmission.
- *Resource-allocation mode 2*: here a decision on sidelink transmission, including the defining of the exact set of resources to be applied during transmission is taken by the transmitting device itself. Thus, this mode is applicable to both the in-coverage and out-of-coverage deployment scenarios.

Sensing and resource selection is the procedure via which a device functioning under resource-allocation mode 2 chooses the set of time/frequency resources for use in sidelink transmission. This selection is based in part on resource reservations announced by other devices. In Release 16, with operation in resource-allocation mode 2, only the so-called *full sensing* is supported. With full sensing, for time critical transmission, the device has to operate with continuous reception so as to have available the sensing information for resource selection should data arrive for sidelink transmission. Full sensing results in measurable energy consumption, but as Release 16 is targeted toward vehicle installed devices this does not present a serious problem.

NR V2X Release 16 specifications support operation in the licensed and unlicensed FR1 (sub-6 GHz) and FR2-1 (mmWave) spectrum. The primary focus, however, is operation in FR1 with the only FR2 specific optimization being support for phase tracking via the PT-RS. The Release 16 supported FR1 operating bands are:

- The licensed 2570–2620 MHz band (n38).
- The unlicensed 5855–5925 MHz "Intelligent Transport Systems" (ITS) band (n47).

The supported channel bandwidths in both of the above two bands are 10, 20, 30, and 40 MHz.

No FR2 supported bands are indicated.

As with NR standard signals, all modulating signals are specially derived sequences. NR-V2X sidelink communication supports subcarrier spacings of 15, 30, and 60 kHz in FR1 bands and subcarrier spacings of 60 and 120 kHz in FR2 bands. Their associations to cyclic prefixs are as for standard NR UL/DL but using only the CP-OFDM waveform, i.e., DFTS-OFDM is not used.

8.11.3 Multiple Transmit/Receive Point (Multi-TRP) Transmission

In downlink multi-TRP operation [2], a serving cell can schedule the UE to receive transmissions from two transmit/receive points (TRPs). With such a scheme, we note that from the receiving UE's viewpoint the transmissions, though originating from two physically separate points (but driven by the same serving cell), still appear to be from a single logical point, but collectively of higher power than that received from any one transmission point. Thus, the multiple transmissions are "transparent" to the device. Device-transparent multi-TRP operation is supported in the NR specification.

In addition to transparent operation, NR specifications also support several "non-transparent" multi-TRP operations.

The transmission simultaneously of different data from two different transmission points to a single device so as to increase the received data rate is supported in

Release 16 and referred to as *Non-Coherent Joint Transmission* (NCJT). Two versions of NCJT are outlined, namely *single-DCI-based NCJT* and *multi-DCI-based NCJT*.

The transmission to a device from two transmission points of the same data but utilizing different time and frequency resources leads to much improved reliability of the received data at the device as here we have a combination of space, time, and frequency diversity. Release 16 supports such communication but with support limited to the PDSCH channel. Given that high reliability is very relevant for URLLC services, it has been referred to as Downlink multi-TRP for URLLC.

8.11.4   NR Positioning

5G positioning is a technology whereby the 5G mobile communications network determines a geographic location of a device by measuring radio signals. In 3GPP release 15, 5G device positioning was provided in the non-standalone mode using 4G positioning reference signals to measure on. This positioning was dependent on support by technologies independent of 3GPP radio access technologies (RAT), in particular on *global navigation satellite systems* (GNSS) technology. There are today several prospective applications, for example manufacturing and the internet-of-things, that require high precision positioning both outdoors and indoors. However, as satellite-based positioning systems have limited indoor coverage and not quite the positioning accuracy sometimes required, NR introduced in Release 16 dedicated 5G positioning signals, measurements, and procedures to provide better positioning support. Typical horizontal positioning accuracy of LTE/Release 15 NR is 20-m outdoors and <5-m indoors. NR Release 16 horizontal accuracy, on the other hand, is targeted to be <10-m outdoor and <3-m indoor for 80% of the UEs, and vertical accuracy expected to be <3-m outdoors and indoors for 80% of the UEs. The targeted requirement for latency is end-to-end latency of <1 s.

Figure 8.16 shows the 5G architecture supporting positioning in a standalone situation [14]. It shows the target UE, a serving gNB, the *access and mobility management function* (AMF) which is located in the 5G core (5GC), and a new entity, the *location management function* (LMF) [15], which is also located in the 5GC. The gNB/AMF interface is the NG control (NG-C) plane interface. The AMF/LMF interface is the NL interface. The LMF is the location computation center and it selects which positioning method [16] will be applied based in part on the gNB and

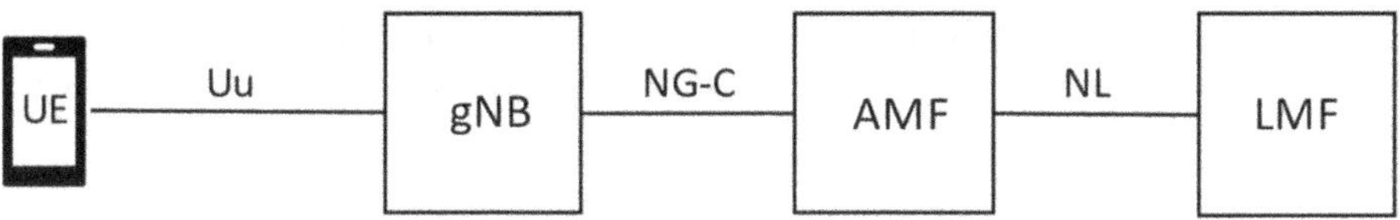

Fig. 8.16 UE positioning architecture applicable to gNB

the UE positioning capabilities. The gNB, as requested, exchanges with the LMS via the AMF the uplink measurements made by itself, and the downlink measurements made by the UE necessary to compute the UE position.

The location process proceeds as follows. The AMF may receive a location request from the associated UE or a request from the gateway mobile location center (GMLC) to initiate the positioning of a target UE, the GMLC being the network entity in the 5G core supporting location services. Also, the AMF may decide on its own to locate a UE that has, for example, placed an emergency call. In all cases, the AMF then sends a location service request to the LMF which processes it and determines the assistance data that is to be transferred to the UE. Included in this data is the positioning method to be used and whether the method is UE-based and or UE-assisted positioning. Finally, the LMF returns the location result back to the AMF which then forward this result to the client that requested the location.

8.12 5G NR Release 17

Release 17, which took place in July 2020, added enhancements and new features to release 16. Among the new features are:

- NR operation in the European 6-GHz licensed band 6425–7125 MHz (n104).
- Operation in the newly defined FR2-2 mmWave range, 52.6–71.0 GHz (See Sect. 8.12.1).
- Adding 1024-QAM as a downlink option in the FR1 band. In Release 15 and 16 FR1 DL modulation order was limited to 256-QAM. This addition increases the maximum DL data rate achievable by 25%.
- Reduced capability (RedCAP) UE, a UE of reduced complexity, smaller physical size, lower than normal power consumption, and lower than normal cost (See Sect. 8.12.2.)
- NR RF Repeaters, network nodes that can be used to extend the coverage created by a base station (See Sect. 8.12.3).
- Non-Terrestrial Network (NTN) operation which involves UE communication with high-altitude platforms, low earth orbit satellites, and geosynchronous orbit satellites, such a network being intended to allow communication in remote areas where terrestrial coverage is absent (See Sect. 8.12.4).
- NR multicast and Broadcast Services, multicast providing the same service and content simultaneously to a dedicated set of UEs whereas broadcast provides the same service and content simultaneously to all UEs in a geographical area.

A comprehensive summary of Release 17 features can be found in [2, 11, 17]. Following we review at a high level, operation in the FR2-2 band, RedCap operation, NR RF repeaters operation, and NTN operation.

8.12.1   FR2-2 Operation

As discussed in Sect. 8.8.3, 3GPP in Release 17 added to its list of approved bands the band 57.0–71.0 GHz (n263) for unlicensed NR operation in the TDD mode. This band falls within the range of FR2-2, offers wide transmission bandwidths ranging from 100 to 2000 MHz and hence the possibility of very high transmission data rates (see Table 8.3). Also in Release 17, new subcarrier spacing options, namely 480 and 960 kHz were added, these options being for application in the FR-2 band only (see Table 8.2). These options allow the effective utilization of the wider transmission bandwidths. We noted in Table 8.2 that with a 120-kHz subcarrier spacing the maximum transmission bandwidth is $120 \times 12 \times 275 = 396$ MHz (recall that the maximum number of supported PRBs is 275) which would be a very poor use of a 1600- or 2000-MHz channel bandwidth. However, a subcarrier spacing of 480/960 kHz provides an efficient use of a 1600/2000-MHz channel. As phase noise increases with operating frequency [16], transmission in the FR2-2 band is especially susceptible to phase noise. However, phase noise induced ICI is reduced by increasing subcarrier spacing. Thus, a further benefit of large subcarrier spacing in this band is the minimization of phase noise induced ICI.

These wider subcarrier spacings raise the question as to whether they create increased susceptibility to signal distortion due to multipath fading (Sect. 3.2.6.3). Were the signal paths to be long this would likely be the case. However, because of the very high-transmission frequencies and accompanying atmospheric absorption, signal paths are likely to be very short, a few hundreds of meters at the most, and this limits any multipath fading to acceptable levels.

8.12.2   RedCap Operation

There are many IoT use cases such as industrial wireless sensors, wearables, and video surveillance where a UE of reduced complexity, smaller physical size, lower than normal power consumption, and lower than normal cost provides an attractive alternative to the normal UE. 3GPP in Release 17 released features to support such a reduced capability (RedCap) device [17, 18]. Some of the key physical layer features of a RedCap UE compared to a normal NR UE are given in Table 8.9.

Among the non-physical layer driven capabilities of RedCap are two UE power saving techniques, namely Extended discontinuous reception (eDRX) in RRC idle/ inactive state, and radio resource management (RRM) measurement relaxation for neighbor cells [17, 18].

In total, 65% cost reduction for FR1 and 50% for FR2 devices is achievable for the simplest RedCap devices over standard UEs [2].

Table 8.9 Comparison of Rel. 17 normal UEs and RedCap UEs

	FR1-Normal UE	FR1-RedCap UE	FR2-Normal UE	FR2-RedCap UE
Device Rx/Tx bandwidth	At least 100 MHz	20 MHz max.	At least 200 MHz	100 MHz max.
Min. antenna configuration	2 or 4 receive branches	1 or 2 receive branches	2 antenna panels, each supporting 4 dual polarized antenna elements	2 antenna panels, each supporting 2 dual polarized antenna elements
Supported DL MIMO layers	2 or 4	1 or 2	1 or 2	1 or 2
Max. DL Mod. order	1024QAM	64QAM mandatory 256QAM optional	256QAM	64QAM mandatory 256QAM optional
Max. UL Mod. order	256QAM	64QAM mandatory 256QAM optional	64QAM (Note 1)	64QAM (Note 1)
Duplex operation	TDD or full-duplex FDD	TDD or half-duplex FDD or full duplex FDD	TDD	TDD
Carrier aggregation	Supported	Not supported	Supported	Not supported
Dual connectivity	Supported	Not supported	Supported	Not supported

Note 1: 256QAM "supported", but 64QAM is max. modulation order measurement channel defined

8.12.3   *NR RF Repeaters*

NR RF repeaters, introduced in Release 17 [17, 19], are network nodes that can be used to extend the coverage created by a base station via (a) the amplifying of signals from the base station and retransmitting them into a poorly covered area and (b) the amplifying of signals from a UE in a poorly covered area and retransmitting them to the serving base station. Other than amplification, repeaters perform no signal processing.

Two repeater types were defined, (a) type 1-C for operation in all FR1 bands except the unlicensed shared spectrum bands, namely bands n46, n96, and n102, and (b) type 2-0 for operation in all FR2-1 bands. FR1 bands include both FDD operation and TDD operation. FR2-1 bands include only TDD bands. When operating in TDD bands, repeaters are synchronized with the base station in whose coverage area they are located so as to follow the UL/DL frame configuration that the base station is using.

Three downlink repeater classes were introduced for type 1-C and type 2-0 repeaters to address different deployment scenarios:

– *Wide Area repeaters*, characterized by requirements derived from Macro-Cell scenarios.
– *Medium Range repeaters*, characterized by requirements derived from Micro-Cell scenarios.
– *Local Area repeaters*, characterized by requirements derived from Pico-Cell scenarios.

Two uplink repeater classes were introduced for type 1-C and type 2-0 repeaters to address different deployment scenarios:

– *Wide Area repeaters*, characterized by requirements derived from Macro- and/or Micro-Cell scenarios.
– *Local Area repeaters*, characterized by requirements derived from Micro- and/or Pico-Cell scenarios.

The spectrum area over which a repeater operates can either be contiguous or non-contiguous.

For type 1-C repeaters, rated output power per repeater class is given in Table 8.10.

For type 2-0 repeaters, rated output power per repeater class is given in Table 8.11.

Table 8.10 Repeater type 1-C rated output power per repeater class

Repeater class	$P_{\text{rated,p,AC}}$	
	DL	UL
Wide area	No upper limit	No upper limit
Medium range	$\leq$38 dBm + X, Note 1	N/A
Local area	$\leq$24 dBm + X, Note 1	$\leq$24 dBm + X, Note 1

$P_{\text{rated,p,AC}}$: rated passband output power per antenna connector
Note 1: $X = 10 \log(\text{ceil}(\text{passband bandwidth}/20 \text{ MHz}))$
Ceil: Round up to the nearest integer

Table 8.11 Repeater type 2-0 rated output power per repeater class

$P_{\text{rated,out,TRP}}$	
DL	UL
No upper limit	No upper limit
No upper limit	N/A
No upper limit	$\leq$35 dBm + X, Note 1

$P_{\text{rated,out,TRP}}$: rated output total radiated power (TRP) declared per RIB
RIB: radiated interface boundary
Note 1: $X = 10 \log(\text{ceil}(\text{passband bandwidth}/100 \text{ MHz}))$
Ceil: round up to the nearest integer

8.12.4   *Non-terrestrial Networks (NTNs)*

Non-Terrestrial Network (NTN) operation involves device communication with *low earth orbit* (LEO) satellites, *geosynchronous orbit* (GSO) satellites, and high-altitude platforms (HAPs), such a network being intended to allow communication in remote areas where terrestrial coverage is absent such as the oceans and very rural areas.

In 3GPPs Release 17, the main emphasis is on communication via GSO and LEO satellites. With GSO satellites, as the name implies, the satellites are at such a height above the earth's surface as to appear stationary relative to the earth's surface, this height being 35,786 km. LEO satellites on the other hand, which normally orbit at a height of about 300–1500 km above the earth's surface, move rapidly relative to the earth's surface. Thus, with communications with such satellites, there is very large and rapidly varying Doppler shift plus the need to regularly switch from communication with one satellite to another as satellites come in and out of line-of-sight. GSO satellite communication is likely for non-delay critical services such as fixed broadband and IoT whereas LEO communication is more likely for low delay requiring services such as two-way voice.

To handle the differences created by satellite communications, NR Release 17 sets out the basic mechanisms necessary to deal with:

– The large propagation delay in the transmissions between the earth-bound device and the satellite, especially with GSO satellites.
– The large doppler shift and need for frequent handovers with LEO satellites.
– Delay variation and Doppler shift variation, especially with LEO satellites.
– High path loss.
– Different propagation channel model compared to terrestrial propagation models.

Rel-17 specified two FDD frequency bands for NTN service:

– 1626.5–1660.5 MHz UL/1525–1559 MHz DL (n255)
– 1980–2010 MHz UL/2170–2200 MHz DL (n256)

Using the above frequencies, it is possible to achieve tens of megabits per second DL speed and round-trip delay in the order of a few tens of milliseconds.

8.13   5G NR Release 18 (5G-Advanced)

Release 18, branded by 3GPP as *5G-Advanced*, was released in June 2024, and adds significant enhancements and new features to the base line created via Releases 15, 16, and 17. Some of these enhancements are:

MIMO enhancements with emphasis on UL MIMO and targeting non-smartphone devices such as FWA, vehicles, and industrial devices.

Enhanced RedCap (eRedCAP): Introduced for IoT use cases with lower peak data rate requirements than RedCap with the following features:

- Peak data rate of 10 Mb/s (is 100 Mb/s in Release 17 RedCap).
- Device bandwidth maximum of 5 MHz (is 20 MHz in Release 17 RedCap).
- Lower device complexity and power consumption than Release 17 RedCap.
- Improved positioning support.
- Lower expected cost than Release 17 RedCap.

Network-Controlled Repeaters (NCRs): These repeaters are an enhancement over conventional repeaters such as supported in Release 17, having the capability to receive and process control information from the network (See Sect. 8.13.1).

IAB enhancements, which includes enhanced support for mobile IAB nodes. Such nodes would be vehicle installed and provide coverage to UEs in or near the vehicle. The served UE would be unaware as to whether the serving cell is a mobile IAB node or a regular gNB but latency may be larger than normal due to the extra hop involved.

NTN evolution, which includes improved coverage for certain services and support for operation above 10 GHz (See Sect. 8.13.2).

Support for smaller than 5-MHz channel bandwidth in FR1, operating in channels of 3–5 MHz bandwidth and using 15-kHz SCS with only normal Cyclic prefix. This support is to accommodate some use cases such as railroad communication where the available channel bandwidths are smaller than 5 MHz.

Comprehensive summaries of Release 18 features can be found in [20–22]. Following we review, at a high-level, Network-Controlled Repeaters and NTN evolution.

8.13.1   Network-Controlled Repeaters

Network-controlled repeaters (NCRs) [23], like non-network-controlled ones, amplify and forward the received signals from the gNB to the UE and from the UE to the gNB with minimum delay. However, they also receive from and transmit to the gNB side control information over a control link (C-link), this link operating on the same frequency as the signal being repeated. Examples of side control

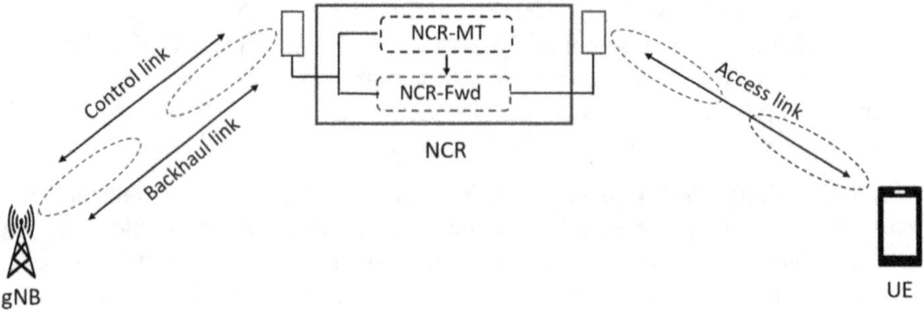

Fig. 8.17 Network-controlled repeater based system

information include beamforming information, power control information to improve energy efficiency, and UL/DL TDD configurations. Particularly important is the beam forming information which allows the repeater to transmit/receive to/ from the UE on a directed beam as well as to transmit/receive to/from the gNB on a directed beam. Figure 8.17 shows an NCR interposed between a gNB and a UE. Between the gNB and the NCR we have a backhaul link for transporting user data and a control link. Between the NCR and the UE we have an access link only. 3GPP models the NCR as consisting of two functional entities, namely the *NCR forwarding* (NCR-Fwd) and the *NCR mobile termination* (NCR-MT).

The NCR-Fwd is the entity that performs the amplifying and forwarding of UL/ DL RF signals between the gNB and the UE via the backhaul and access links. The functioning of the NCR-Fwd is controlled via the control information received from the gNB via the C-link.

The NCR-MT is the entity that communicates with the gNB via the C-link to enable the exchange of information. The C-link is based on the NR Uu interface.

Release 18 only supports single-hop repeating, that is, no additional repeating following the first repeater.

8.13.2 NTN Evolution

Release 18 NTN evolution includes improved coverage for low-data rate and VoIP services via smartphones, low-data rate being available via all satellite orbit types while VoIP only available via LEO satellites due to low latency requirements.

Evolution also includes support for operation above 10 GHz with VSAT (very small aperture terminal) small dish antennas systems. Specifically, the following three FDD frequency bands are supported:

– 17.7–20.2 GHz UL/27.5–28.35 GHz DL (n510)
– 17.2–20.2 GHz UL/28.35–30.0 GHz DL (n511)
– 17.2–20.2 GHz UL/27.50–30.0 GHz DL (n512)

Using the above frequencies, it should be possible to achieve DL speed of up to about 100 Mb/s with fixed VSAT devices. These bands are not applicable to smartphones and IoT devices because of the large path loss resulting from the high operating frequency coupled with the low device antenna gain.

8.14   Summary

In this chapter, key elements of 5G/5G-Advanced were reviewed with particular emphasis on enhanced mobile broadband (eMBB) as this is the feature that most impacts smartphone communication. First, the main architecture options for connection to the core network were reviewed followed by the *Radio Access Network* (RAN) protocol architecture. The RAN is responsible for all radio-related functions such as coding, modulation, HARQ operation, physical transmission, scheduling, etc. Following this, study within the RAN was narrowed to the physical layer as specified in 3GPP Release 15, the first 5G release, with emphasis on how the user data physical channels are structured. How maximum user data rates and low latency are achieved was described, 5G operating frequency spectrum was reviewed, and some typical base station and UE parameters presented. Next, we addressed certain key features introduced in Release 16, Release 17, and Release 18, i.e., 5G-Advanced. In earlier chapters, key 5G physical layer technologies were introduced. In this chapter, where and how in NR these technologies are applied have been shown. The NR physical layer is the most complex and most flexible point-to-multipoint radio access system introduced to date, offering the potential of unprecedented data rates and low latency. It is largely due to the technologies introduced in earlier chapters of this text and the way in which they have been applied, as shown in this chapter, that has made this progress possible.

The material presented in this chapter has been at the introductory level. For those readers seeking more detail than presented here, see [2, 16].

References

1. 3GPP TR 38.912 (2018) Study on new radio access technology, Rel. 15, version 15.0.0
2. Dahlman E et al (2023) 5G/5G-advanced: the new generation wireless access technology, 3rd edn. Academic Press, London
3. Holma H et al (eds) (2020) 5G technology: 3GPP new radio. John Wiley & Sons Ltd., Hoboken
4. 3GPP (2017) TR 38.801: study on radio access technology: radio access architecture and interfaces, Sophia Antipolis, France
5. Ahmadi S (2019) 5G NR: architecture, technology, implementation, and operation of 3GPP new radio standards. Academic Press, London
6. 3GPP TS 38.306 (2022-05) NR; user equipment (UE) radio access capabilities, Rel. 17, version 17.0.0
7. 3GPP TS 38.104 (2022-8) NR; base station (BS) radio transmission and reception, Rel. 17, version 17.6.0

8. 3GPP TS 38.101-1 (2022-05) NR; user equipment (UE) radio transmission and reception; part 1; range 1 standalone, Rel.17, version 17.5.0
9. 3GPP TS 38.101-2 (2022-08) NR; user equipment (UE) radio transmission and reception; part 2; range 2 standalone, Rel. 17, version 17.6.0
10. 3GPP TR 21.916 (2021): Release 16 description; summary of Rel-16 work items, Rel. 16, version 16.0.1
11. 5G Americas (2021) 3GPP Releases 16 & 17 & Beyond, Bellevue, Washington, U.S.A
12. 3GPP (2018) TR 38.874: study on integrated access and backhaul, Sophia Antipolis, France
13. 3GPP (2020) TR 37.985, version 16.0.0: overall description of radio access network (RAN) aspects for vehicle-to-everything (V2X) based on LTE and NR, Sophia Antipolis, France
14. 3GPP (2020) TS 38.305, version 16.1.0: stage 2 functional specification of user equipment (UE) positioning in NG-NR, Sophia Antipolis, France
15. 3GPP (2020) TS 29.572 version 16.3.0: location management services; stage 3, Sophia Antipolis, France
16. Morais DH (2024) Key 5G/5G-advanced physical layer technologies, 3rd edn. Springer, Cham
17. 3GPP TR 21.917 (2023) Release 17 description; summary of Rel-17 work items, Rel. 17, version 17.0.1, Sophia Antipolis, France
18. Veedu SNK et al (2022) Towards smaller and lower-cost devices with longer battery life: an overview of 3GPP Release 17 RedCap. IEEE Commun Stand Mag 6(3):84–90
19. 3GPP (2022) TS 38.106, version 17.1.0: NR repeater radio transmission and reception, Sophia Antipolis, France
20. Chen W et al (2023) 5G-advanced towards 6G: past, present, and future. IEEE J Sel Areas Commun 41(6):1592
21. 5G Americas (2022) Becoming 5G-advanced: the 3GPP 2025 roadmap. Bellevue, Washington
22. 3GPP TR 21.918 (2024) Release 18 description: summary of Rel-18 work items, Rel. 18, version 1.1.0, Sophia Antipolis, France
23. 3GPP TR 38.867 (2022) Study on NR network-controlled repeaters, (Release 18), version V0.1.0, Sophia Antipolis, France

Chapter 9
Wi-Fi 6/6E and Wi-Fi 7 Overview

9.1 Introduction

As indicated in Sect. 1.3, Wi-Fi is officially a collection of wireless network protocols, based on the IEEE 802.11 series of standards and used for local area networking, but more commonly for local Internet access, which is how it is used with smartphones. It has become a ubiquitous technology worldwide, being the first choice of progressively more users to connect with the Internet. Wi-Fi today forms an integral part of smartphones. Though smartphone Internet connection is available via mobile access, depending on location, such access is not always at high speed, sometimes not available, and in some instances when available costly. With Wi-Fi capability, if close to a Wi-Fi access point, a smartphone can typically access the Internet at very high speeds and no cost.

Wi-Fi 6/6E (802.11ax High Efficiency) [1], the release following Wi-Fi 5 (802.11ac Very High Throughput) [2], is designed to facilitate high-density wireless access and high-capacity wireless services, in locations such as high-density sports arenas, indoor high-density offices, large-scale outdoor public venues, etc. Compared to Wi-Fi 5, Wi-Fi 6/6E employs technologies such as OFDMA, downlink and uplink multi-user MIMO, 1024-QAM modulation, etc., that allow it to achieve an approximate fourfold increase in average user throughput, to increase the number of concurrent users more than threefold and to increase the maximum theoretical data rate by 38%. In the following sections, we first examine, for Wi-Fi in general, Internet access network architecture and protocol architecture. This is followed by a study of the RF spectrum supported by both Wi-Fi 6/6E and Wi-Fi 7 (802.11be Extremely High Throughput). Next, for Wi-Fi 6/6E, we review the physical layer and key technologies employed therein, and some new procedures for improving medium utilization and battery efficiency. We then examine the key

© The Author(s), under exclusive license to Springer Nature
Switzerland AG 2025

D. H. Morais, *5G/5G-Advanced, Wi-Fi 6/7, and Bluetooth 5/6*,
https://doi.org/10.1007/978-3-031-82830-0_9

features of Wi-Fi 7 (802.11be Extremely High Throughput). Wi-Fi 7 introduces technologies that improve reliability, reduce latency, while continuing to increase data rates. Finally, a comparison between Wi-Fi 7, Wi-Fi 6/6E, and Wi-Fi 5 is presented.

9.2   Internet Access Network Architecture

Following are the main 802.11 architecture components:

- **Stations (STA)**: All the Wi-Fi devices in communication with each other are called *Stations*.
- **Access Point (AP) Station**: An AP-STA is the equivalent of a mobile base station, being a central unit that communicates with one or several other units. It is normally referred to just an Access Point.
- **Non-Access Point Station**: These are the stations that the AP communicates with and are normally referred to as *clients* or sometimes just Stations. Examples are smartphones, computers, and printers.
- **Basic Service Set (BSS)**: A BSS is a group of stations communicating with each other at the physical level. A BSS can be an Infrastructure BSS or an Independent BSS.
- **Infrastructure BSS**: When clients communicate with each other through an AP, the set is called an *Infrastructure BSS*. This configuration is commonly used in homes, offices, and hotspot network installations. A smartphone will typically acquire Wi-Fi access by joining an infrastructure BSS.
- **Independent BSS**: When clients communicate directly with one another in a peer-to-peer, ad-hoc fashion without an AP, the set is called an *Independent BSS*. An Independent BSS has no connection to an external network and is not a popular configuration.
- **Extended Service Set (ESS)**: This is a set of multiple BSSs where the APs are all connected to a common *Distribution System* (DS).

Figure 9.1 shows an Infrastructure BSS with an AP communicating with three clients, namely a smartphone, a computer, and a TV.

Figure 9.2 shows a typical structure for connecting an AP to the Internet. Following is a description of the individual units shown:

- The *modem* is the gateway to the Internet. On its left, we see a coaxial or fiber-optic cable (in the past this connection could also have been via old fashioned twisted pair telephone cable, but this practice is now just about obsolete due to the low bit rates afforded). This coaxial or optical cable originates with the Internet provider and communicates via electrically or optically modulated analog signals that convey IP data. In the inbound direction, the modem translates these signals into Ethernet link layer ones (Sect. 2.2.5). In the outbound direc-

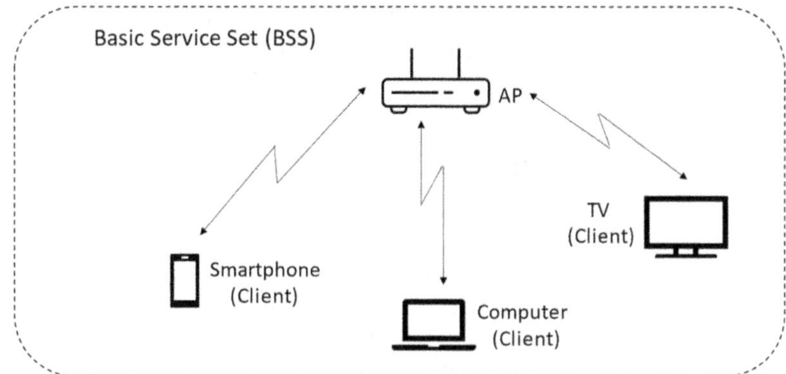

Fig. 9.1 A typical infrastructure basic service set

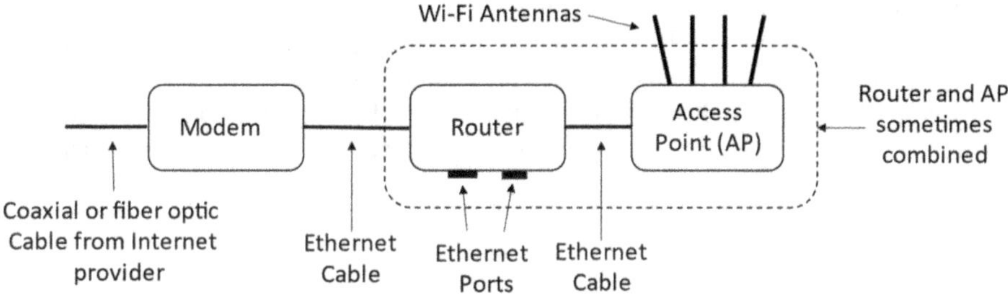

Fig. 9.2 A typical access point connection to the Internet

tion, it translates Ethernet link layer signals to modulated analog electrical or optical ones. A standalone modem normally has only one Ethernet port, contains only one IP address, and is unable to send/receive data to/from multiple devices simultaneously. To accomplish this, a router is required. To the right of the modem a connection to a router is shown.

- The **router**, as its name implies, routes traffic to and collates traffic from one or more devices. It gives each device it is communicating with its own IP internal address which is a subaddress of the router address. All its communication is via Ethernet signals. For a wired connection to a computer, for example, an Ethernet cable would be connected between the router and the computer. In the figure a connection between the router and an Access Point (AP) is shown.
- The **Access Point** is what facilitates Wi-Fi wireless communication with clients and its connection with the router is an Ethernet one. It can be a standalone device but often is combined together physically with a router.

Though Fig. 9.2 shows only one AP, it is now common, in order to cover large areas with lots of dead zones when covered by only one AP, to provide coverage via multiple specialized APs configured in what's known as a *mesh network* or more formally a *mesh basic service set* (MBSS). With such a network, all APs within range of each other are wirelessly connected directly to each other.

9.3   Protocol Architecture for Internet Traffic

The IEEE 802.11 standards cover Layer 1, the Physical Layer, and Layer 2, the Data Link Layer, of the IEEE 802 Reference model. Figure 9.3 shows these layers as part of a protocol stack that supports Internet traffic. Layers 3, 4, and 5 have been covered in Chap. 2. Our attention here, therefore, will be on Layers 1 and 2.

Before proceeding further, the definition of some commonly used terms in describing the various protocols is in order:

- A *data unit* is the basic unit exchanged between different layers of a protocol stack.
- A *service data unit* (SDU) is a data unit passed by a layer above to the current layer for transmission using services of the current layer.
- A *protocol data unit* (PDU) is a data unit created by the current layer via the adding of a header (and sometimes a trailer) to the received SDU prior to transportation to the layer below. The added header describes the processing carried out by the current layer.

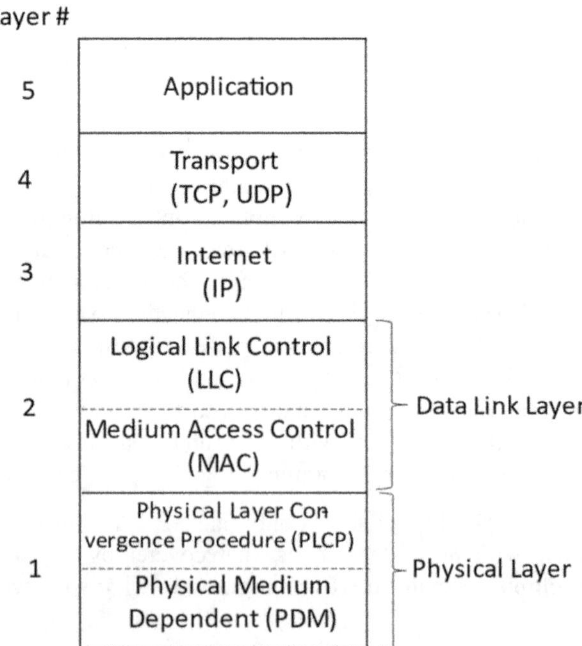

Fig. 9.3 802.11 protocol stack for Internet data transmission

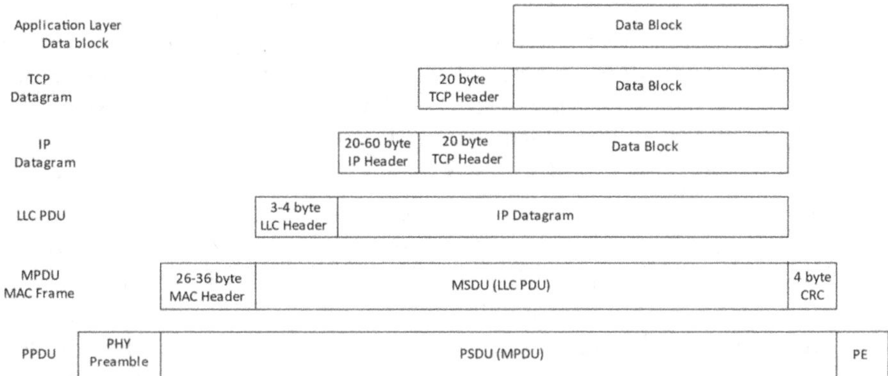

Fig. 9.4 Data flow through protocol stack

Figure 9.4 shows the flow of data and the appending of headers and a trailer through the protocol stack where the above definitions are applied.

9.3.1   Data Link Layer

As can be seen in Fig. 9.3, the Data Link layer consists of two sublayers, namely the *Logical Link Control* (LLC) sublayer and the *Medium Access Control* (MAC) sublayer.

The LLC Sublayer The LLC sublayer is tasked with facilitating the transmission of a link-level PDU between two stations. The LLC sublayer used by 802.11 (802.2 LLC) supports the transmission of several Network Layer (Layer 3) protocols, including IP, within a multipoint network over the same wireless medium. In the direction of downward data flow, in the case of IP transmission, IP datagrams are passed down to the LLC sublayer, which attaches a header to create an LLC protocol data unit (LLC PDU). The header contains a one-byte destination address field (Destination Service Access Point (DSAP)) and a one-byte source address field (Source Service Access Point (SSAP)) and specifies that the transported packets are IP ones. It also contains a one- or two-byte control field which can be used for hop-to-hop flow and error control. Note, however, that, for TCP/IP protocols running above, this flow and error control is not used as it is taken care of in the TCP protocol. At the receive end, on receiving a frame from the physical layer below, the LLC determines from the header that the Level 3 protocol type is IP and thus hands over the datagram to the IP layer above. Often the LLC header as described above is expanded by inserting just before the IP datagram a five-byte field called a Subnetwork Access Protocol (SNAP) extension which bestows certain advantages.

The MAC Sublayer Below the LLC sublayer is the MAC sublayer. It is designed to support multiple users on a shared RF medium by having the transmitter sense the availability of the medium before accessing it. In the direction of downward data flow, higher level data in the form of an LLC PDU is passed down to the MAC sublayer, which attaches a header and a trailer to create a MAC protocol data unit (MPDU) also referred to as a MAC frame. Its functions include:

- As the name implies, controlling access to the transmission medium. The protocol employed is called Carrier Sense Multiple Access/Collision Avoidance (CSMA/CA). It requires each station to first check on the state of the medium before initiating a transmission and then only doing so if the medium appears to be free. This minimizes the potential of a collision and hence loss of data occurring as a result of having two or more stations transmitting simultaneously.
- On transmission, assembling data into a frame with destination address, source address, and error detection (CRC) fields.
- On reception, disassembling received frame and performing address recognition and error detection.
- On reception, via the CRC attachment, detecting errors that may have occur in the physical layer.

9.3.2   Physical Layer

As can be seen in Fig. 9.3, the Physical layer, like the Data Link layer, consists of two sublayers, namely the *Physical Layer Convergence Procedure* (PLCP) sublayer and the *Physical Medium Dependent* (PMD) sublayer.

The Physical Layer Convergence Procedure (PLCP) Sublayer In the downward direction of data flow, higher level data in the form of an MPUD is passed down to the PLCP sublayer where it is termed a PLCP SDU (PSDU). Here a header and a trailer (Packet Extension) are attached to create a PLCP protocol data unit (PPDU) also referred to as a PLCP frame. The PPDU is of a framing format appropriate for sending and receiving user and management data between two or more stations using the associated PMD sublayer below. The added header fields contain information needed by the PMD sublayer transmitter and receiver as well as a PHY-specific preamble to aid the receiver in the synchronization of incoming transmissions. The synchronization capability is required because the PPDU frame structure supports asynchronous transfer of PSDUs between stations resulting in the need in the receiving station to synchronize to each incoming frame. In the receive mode, it strips the header and trailer from incoming frames delivered from the PMD sublayer below and passes on the remainder (MPUDs) to the MAC sublayer above. 802.11ax specifies four different formats of the PPDU frame. These formats have many components, and they will be discussed in Sects. 9.5.6. and 9.6.8.

The Physical Medium Dependent (PMD) Sublayer This sublayer, under the direction of the PLCP sublayer, provides the transmission and reception of physical layer user and management data (PPDU frames) across a wireless medium between two or more stations, interfacing directly with the wireless medium. Its functions include coding/decoding and modulation/demodulation of the frame transmissions, and signal transmission and reception over the air via antennas. Also, it incorporates a clear channel assessment (CCA) function to appraise the RF spectrum and indicate to the MAC whether or not a signal has been detected. The medium must be clear before a transmission can commence.

9.4   Spectrum for Wi-Fi 6/6E and Wi-Fi 7

Wi-Fi links operate only in unlicensed frequency bands. In such bands frequency planning before operation is permitted is avoided. However, with multiple simultaneous users in such an environment interference is almost a given unless special procedures are in place. Such interference is avoided by requiring that any new access of the spectrum is only permitted if at the time of required access the spectrum is free.

The 2019 release of 802.11ax (Wi-Fi 6) supports operation in the 2.4- and 5-GHz bands. The 2021 release of updated 802.11ax (Wi-Fi 6E) added support for operation in the 6-GHz band in addition to support for 2.4- and 5-GHz bands. The 2024 expected release of 802.11be will support the same bands as Wi-Fi 6/6E, i.e., 2.4, 5, and 6 GHz.

9.4.1   The 2.4-GHz Band

The 2.4-GHz band covers from 2400 to 2495 MHz. Table 9.1 provides the band channelized into 14 overlapping 22-MHz channels (each channel 20-MHz bandwidth plus 1 MHz on either end for inter-channel spacing), centered 5 MHz apart except for Channel 14 which is 12 MHz above Channel 13. In North America, only channels 1–11 are authorized. In Europe and Asia Pacific excluding Japan channels 1–13 are authorized. In Japan, channels 1–14 are authorized, but Channel 14 is only authorized for 802.11b.

With 22-MHz channels only 3 non-overlapping channels are available, namely channels 1, 6, and 11. With 20-MHz channels, 4 non-overlapping channels are available, namely channels 1, 5, 9, and 13. One 40-MHz channel is possible by aggregating two 20-MHz channels. This approach, however, is typically avoided, due to greater interference potential with other users and the more readily available 40-MHz channels in the 5-GHz band.

Table 9.1 2.4-GHz band channelization

Channel	Frequency (MHZ)		
	F-11	Center (F)	F + 11
1	2401	2412	2423
2	2406	2417	2428
3	2411	2422	2433
4	2416	2427	2438
5	2421	2432	2443
6	2426	2437	2448
7	2431	2442	2453
8	2436	2447	2458
9	2441	2452	2463
10	2446	2457	2468
11	2451	2462	2473
12	2456	2467	2478
13	2461	2472	2483
14	2473	2484	2495

Note: Channel 14 authorized in Japan only for 802.11b only

9.4.2   The 5-GHz Band

The 5-GHz Unlicensed National Information Infrastructure (UNII) band defined by the U. S. Federal Communications Commission (FCC) covers the spectrum from 5170 to 5835 MHz. As with the 2.4-GHz band, each designated channel is centered 5 MHz apart. Table 9.2 provides the resulting 25 non-overlapping 20-MHz channels and the frequency band breakdown by UNII designation. With channel aggregation, this band supports 12 40-MHz channels, 6 80-MHz channels, and 2 160-MHz channels. It is currently the main band for very high-speed Wi-Fi but over time this position will be shared with the newly authorized 6 GHz band.

9.4.3   The 6-GHz Band

Wi-Fi 6E and Wi-Fi 7 operation is now supported in the 6-GHz UNII band. In this band, the spectrum extends from 5.925 to 7.125 GHz. With this 1200 MHz, it supports 59 20-MHz channels, 29 40-MHz channels, 14 80-MHz channels, 7 160-MHz channels, and 3 320-MHz channels, thus adding considerably to the Wi-Fi spectrum available in the 5-GHZ band.

In 2021, the United States decided to allow unlicensed use in the 6-GHz band. See Sect. 8.8.2 for more information on this. Also in 2021, the European Union (EU) took its first step toward opening up the entire 6-GHz band to unlicensed operation by releasing 480 MHz (5945–6425 MHz) for this class of service. It is

Table 9.2 5-GHz band channelization

Frequency band	Channel number	Center frequency (MHz)
5150–5250 MHz UNII-1	36	5180
	40	5200
	44	5220
	48	5240
5250–5350 UNII-2	52	5260
	56	5280
	60	5300
	64	5320
5470–5725 UNII-2 extended	100	5500
	104	5520
	108	5540
	112	5560
	116	5580
	120	5600
	124	5620
	128	5640
	132	5660
	136	5680
	140	5700
	144	5720
5725–5850 UNII-3	149	5745
	153	5765
	157	5785
	161	5805
	165	5825

anticipated that in time, with few exceptions, the entire 6-GHz band will be open to unlicensed service worldwide.

9.4.4   *Performance Differences Between 2.4-, 5-, and 6-GHz Wi-Fi Systems*

The indoor range of Wi-Fi systems decreases as frequency increases because higher frequencies suffer greater loss when traversing solid objects such as walls and floors and ceilings. However, higher frequencies such as those in the 5- and 6-GHz frequency bands allow data to be transmitted at higher rates as they provide wider channel bandwidths. Thus:

– The 2.4-GHz band provides to most coverage but transmits data at slower speeds than the 5- and 6-GHZ bands as even though in theory it supports one 40-MHz channel, channel bandwidth available is essentially limited to 20 MHz.

– The 5-GHz band provides less coverage than the 2.4-GHz band but can transmit data at faster speeds via channel aggregation, having 12, 6, and 2 40-, 80-, and 160-MHz channels, respectively. Furthermore, this band tends to suffer less interference than the 2.4-GHz band as fewer devices use it.
– The 6-GHz band provides slightly less coverage than the 5-GHz band but can transmit data at similar rates, having 59, 29, and 14 40-, 80-, and 160-MHz channels, respectively. Depending on location, a number of channels, however, may be unavailable due to the presence of fixed licensed systems.

9.5   Wi-Fi 6/6E Key Technologies

Wi-Fi 6 (802.11ax) was released by the IEEE in 2019 and supports operation in the 2.4- and 5-GHz bands. In 2020 support for operation in the 6-GHz band was added and the nomenclature changed to Wi-Fi 6E. In this section key technologies applied in Wi-Fi 6/6E so as to facilitate its performance are introduced. They are largely evolutionary, building on those used in earlier releases. The main drivers are the unending need for higher and higher data rates and the ability to provide acceptable service to more and more users in localized high population density environments such as sports stadiums. Frequent reference will be made to the technology fundamentals outlined in earlier chapters. Informative material on Wi-Fi 6/6E can be found in [3, 4].

9.5.1   OFDMA

Prior to 802.11ax, all releasers except 802.11b used OFDM (Sect. 6.2) for data transmission. As a result, users were differentiated by time segments where, in a given segment, only one user occupied all subcarriers on one or more OFDM symbols. 802.11ax, however, uses OFDMA (Sect. 6.3) where, as was shown in Fig. 6.8 and reshown here in Fig. 9.5, a given time segment can be shared by multiple users. Such an arrangement is a more efficient data transmission mode allowing optimum use of the spectrum when several users are communicating at the same time. For a given number of subcarriers and a given modulation, the maximum data rate available to a given user, should the total capacity be assigned to that user, is the same for OFDM and OFDMA. However, in a high user situation, where in the case of OFDM each user would contend inefficiently for its turn to use the channel, OFDMA gives then access simultaneously, albeit with less subcarriers each, thus improving the average throughput per user.

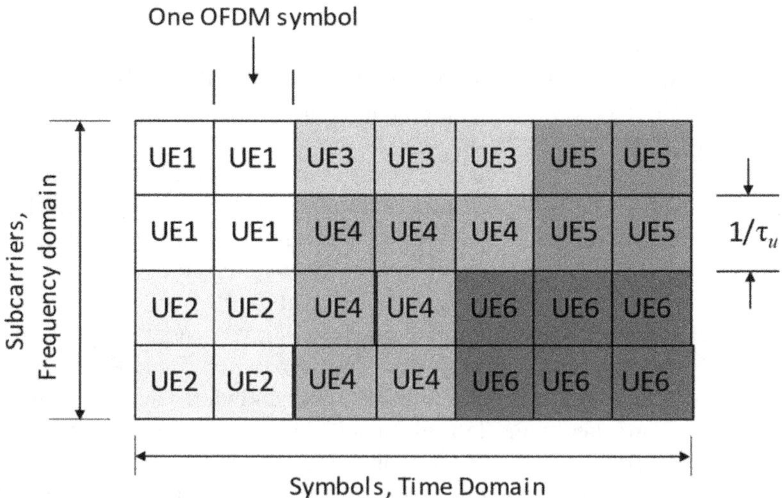

Fig. 9.5 OFDMA scheduling example

Table 9.3 Symbol lengths and guard intervals supported by 802.11

802.11 release	IFFT symbol length, T_S (µs)	Guard interval, T_{GI} (µs)	Transmission efficiency $[T_S/(T_S + T_{GI})] \times 100\%$
Before 802.11ax	3.2	0.4, 0.8	89% for $T_{GI} = 0.4$
802.11ax	12.8	0.8, 1.6, 3.2	94% for $T_{GI} = 0.8$

9.5.1.1   Subcarrier Spacing and Symbol Length

Table 9.3 compares the symbol lengths and *guard intervals* (GIs) supported in earlier releases of 802.11 using OFDM with the 802.11ax release as well as the resulting time domain transmission efficiency.

In earlier releases using OFDM, the subcarrier spacing for both data and control transmissions was 312.5 KHz. This resulted in the IFFT OFDM symbol of length 3.2 µs (1/0.3125 MHz) shown. For 802.11ax, the subcarrier spacing for data transmission has been reduced by a factor of 4–78.125 KHz. This results in the IFFT OFDM symbol of length 12.8 µs (1/0.078125 MHz) shown. Note, however, that for 802.11ax PPDU preamble fields, as will be shown in Sect. 9.5.6, IFFT OFDM symbol lengths are 3.2 µs with one exception where they may be 6.4 or 12.8 µs.

We note from Table 9.3 that 802.11ax affords a higher time domain transmission efficiency when using the minimum guard intervals supported compared to earlier releases. For outdoor transmission or any environment where there is significant delay spread due to reflected signals, the longer GIs supported in 802.11ax allow better symbol recovery than the earlier releases.

9.5.1.2   Resource Units (RUs)

To enable multiple users to share individual OFDM symbols, 802.11ax channels can be divided into subchannels and individual users can be assigned one or more subchannels. These subchannels are called *Resource Units* (RUs). The smallest RU contains 26 subcarriers and occupies approximately 2 MHz. An RU can contain 26/52/106/242/484/996/2x996 subcarriers. Figure 9.6 shows the number of RUs of varying sizes possible in a 20-MHz Channel and the number of subcarriers per RU. If structured to have 9 26 channel RUs, then up to nine simultaneous users could be accommodated. 802.11ax provides support for 20-, 40-, 80-, and 160-MHz contiguous channel widths, and support for 80 + 80-MHz noncontiguous channel widths, depending on the frequency band and capability. Table 9.4 provides the number of RUs supported in the different channel bandwidths defined in 802.11ax.

A 20-MHz channel contains 256 subcarriers. However, we note that the number of subcarriers carrying data is less than 256. For example, if we have nine 26-subcarrier, 2-MHz RUs, we have 216 data carrying subcarriers. This difference is because not all subcarriers carry data. Some are used as guard spaces to minimize interference from adjacent channels, some are used as pilot subcarriers to aid in channel estimation and synchronization, and some are the so-called DC subcarriers, where DC stands for direct current whose frequency or frequencies generally coincide with the channel center frequency.

Note that:

- For a 20-MHz channel carrying one 242-subcarrier RU, there are 234 data subcarriers.
- For a 40-MHz channel carrying one 484-subcarrier RU, there are 468 data subcarriers.

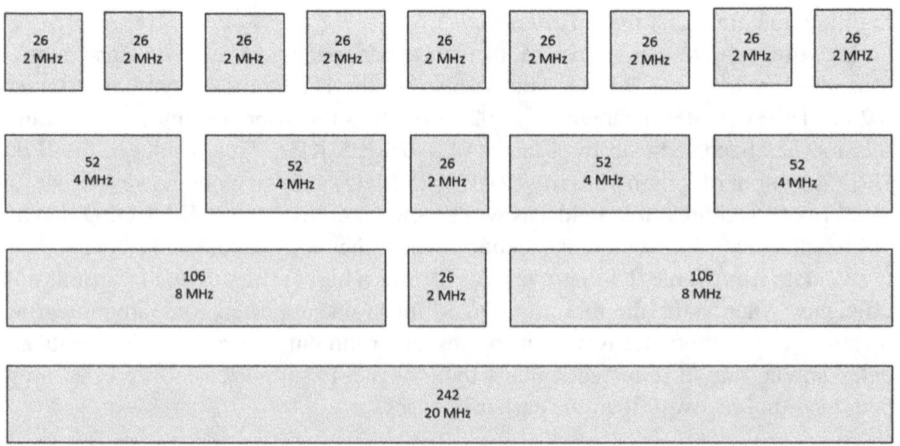

Fig. 9.6 Number of RUs in a 20-MHz Channel

Table 9.4 Maximum number of RUs supported by different channel bandwidths

RU size	20 MHz	40 MHz	80 MHz	160 MHz
26 subcarriers	9	18	37	74
52 subcarriers	4	8	16	32
106 subcarriers	2	4	8	16
242 subcarriers	1	2	4	8
484 subcarriers	N/A	1	2	4
996 subcarriers	N/A	N/A	1	2
2x 996 subcarriers	N/A	N/A	N/A	1

- For an 80-MHz channel carrying one 996-subcarrier RU, there are 980 data subcarriers.
- For a 160-MHz channel carrying one 1992-subcarrier RU, there are 1960 data subcarriers.

Also note that, via carrier aggregation, 80 + 80-MHz channels are supported in the 5- and 6-GHz bands.

9.5.1.3   Downlink OFDMA (DL-OFDMA)

Downlink (DL) 802.11ax OFDMA (Sect. 6.3) multi-user operation allows simultaneous traffic from an AP to one or more clients. This operation can be carried out simultaneously with DL MU-MIMO (Sect. 7.3.4). In DL-OFDMA communication the AP knows what data is to be transmitted. Its job is to dispatch this data in a form that is defined by information such as RU placement, modulation, coding, and client identification. This information is contained in the HE-SIG-B field of the DL Multi-User PPDU frame (Sect. 9.5.6). When a client receives the PPDU packet, it can decode it to determine if it is one of the intended recipients and if so, what are its designated RUs and coding schemes and thus demodulate its intended data.

The AP has the flexibility to allocate any combination of RUs. It decides how the mix and match RU sizes per client at each transmission opportunity (TXOP) in order to dynamically adapt to client data needs. The AP will typically assign larger RUs to clients requiring greater data transfer such as video streaming and smaller RUs to those requiring lesser data transfer such as voice and email. By distributing the available bandwidth among clients, DL-OFDMA uses the channel more efficiently relative to OFDM, assigning to each client only as much of the bandwidth as needed on a per PPDU frame basis.

The DL-OFDMA process proceeds as follows:

- The AP sends a request-to-send frame to clients. This frame contains a RU assignment per client and a timer to inform the clients how long the exchange will take.
- The clients send a clear-to-send response simultaneously using their assigned RUs.
- The AP transmits each client's data simultaneously using the assigned RUs.

- The AP sends an acknowledgment request to confirm that each client successfully received its data frames.
- If the frames were received successfully, the clients reply simultaneously with an
  acknowledgment.

In DL-OFDMA, STBC (Sect. 7.2.1) is applied when there are more transmit than
receive antennas supporting one spatial stream (one data stream flow) and only if
DCM (Sect. 9.5.4.1) is not applied. Furthermore, STBC is used in all RUs or not
used in any of the RUs.

9.5.1.4   Uplink OFDMA (UL-OFDMA)

Uplink (UL) 802.11ax OFDMA (Sect. 6.3) multi-user operation allows an AP to
solicit simultaneous immediate response frames from one or more of the APs clients.
UL-OFDMA is similar to DL-OFDMA. Here multiple clients transmit simultaneously but with each using different RUs within the same channel. UL-OFDMA is a
more difficult process to coordinate than DL-OFDMA as multiple clients must be
coordinated. The AP must determine the best grouping of clients and then signal to
each exactly when and how it should transmit and on which RUs. It signals these
instructions via a so-called Trigger Frame. Information transmitted by the Trigger
Frame in addition to RU allocation includes the number of spatial streams, the duration of the UL-PPDU, and control information such as the client station transmit power.
   The UL-OFDMA process proceeds as follows:

- The 802.11ax AP polls the clients to determine how much buffered data they are
  ready to send.
- The clients reply with a buffered status report.
- The AP sends a request-to-send frame to all clients that contains RU assignments
  and a timer to inform clients how long the exchange will take.
- The clients send a clear-to-send response simultaneously using their assigned RUs.
- The AP sends a final trigger frame to coordinate each client's transmission.
- The clients transmit their data simultaneously.
- If data frames from the clients successfully received, the AP responds with an
  acknowledgment.

As with DL-OFDMA, in UL-OFDMA, STBC is applied when there is only one
spatial stream and only if DCM (Sect. 9.5.4.1) is not applied. Furthermore, STBC is
used in all RUs or not used in any of the RUs.

9.5.2   *Multi-user MIMO (MU-MIMO)*

The fundamentals of multi-user MIMO (MU-MIMO) operation have been presented in Sect. 7.3.4. The 802.11ax specification supports both downlink and uplink
MU-MIMO and up to eight antennas are accommodated. Thus, it can

simultaneously transmit in the downlink data from an AP to a maximum of eight clients and in the uplink data from a maximum of eight clients to the AP. It can thus increase the overall system capacity by a factor of 8 in both the uplink and downlink.

Though the theoretical system throughput gain from MU-MIMO operation can be considerable, it is important to remember that acceptable operation is only possible where the propagation characteristics are such as to allow the AP to determine that the transmission optimized for one client will not be such as to significantly interfere with another and vice versa. For example, if two clients are located very close to each other, then it becomes almost impossible to create spatial streams to each that are sufficiently uncorrelated from each other.

9.5.2.1   Downlink MU-MIMO

As indicated above, 802.11ax DL MU-MIMO operation (Sect. 7.3.4) allows simultaneous traffic from an AP to from one to a maximum of eight clients. This operation can be carried out simultaneously with DL-OFDMA. Though up to eight antennas are supported, clients typically have two antennas only and usually no more than four. With eight antennas on the AP, eight spatial streams can be created. However, each client can only decode a number of spatial streams equal to its number of antennas. Thus, an eight antenna AP can serve, for example, eight clients with one spatial stream each, or four clients with two spatial streams each. Figure 9.7 depicts a simplified presentation of a downlink multi-user Wi-Fi system where the AP has four antennas and is communicating with four clients, each with one antenna. From an operational point of view, the AP views this situation as simply a 4×4 MIMO system.

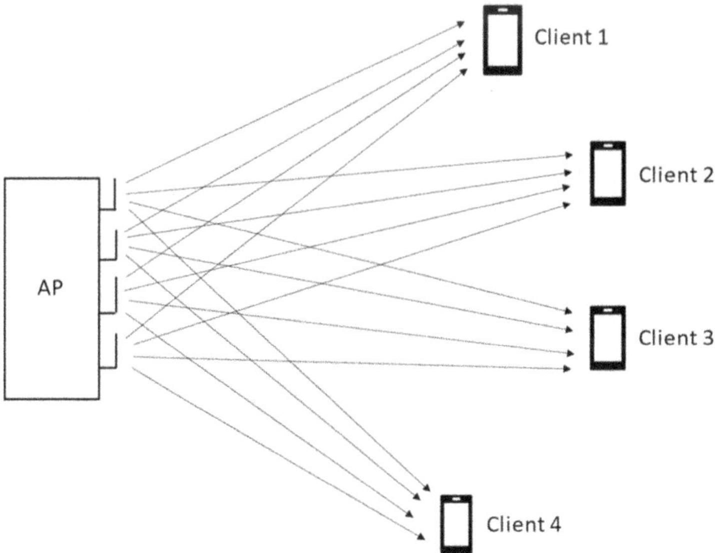

Fig. 9.7 A DL MU-MIMO system

9.5.2.2   Uplink MU-MIMO

Also as indicated above, 802.11ax UL MU-MIMO operation (Sect. 7.3.4) allows an AP to solicit simultaneous immediate response frames from one up to a maximum of eight APs clients. This operation can be carried out simultaneously with UL-OFDMA. For a client with four antennas, for example, using UL-MIMO, it is possible to increase that client's uplink capacity by a factor of 4. Figure 9.8 depicts a simplified presentation of an uplink multi-user Wi-Fi system where the AP has four antennas and is communicating with four clients, each with one antenna. As with the downlink, from an operational point of view, the AP views this situation as simply a 4 × 4 MIMO system.

9.5.3   SU-MIMO and DL MU-MIMO Beamforming

802.11ax supports an adaptive beamforming procedure similar that supported in 802.11 ac on DL and UL SU-MIMO and DL MU-MIMO. Adaptive beamforming is described in Sect. 7.3.2. With SU-MIMO beamforming, all the spatial streams in the transmitted signal are focused on a single station. With DL MU-MIMO, subsets of the spatial streams are focused on different stations. Under this procedure the transmit beamformer initiates a channel sounding procedure. The receiver measures the channel between the transmitter and receiver and feeds back the

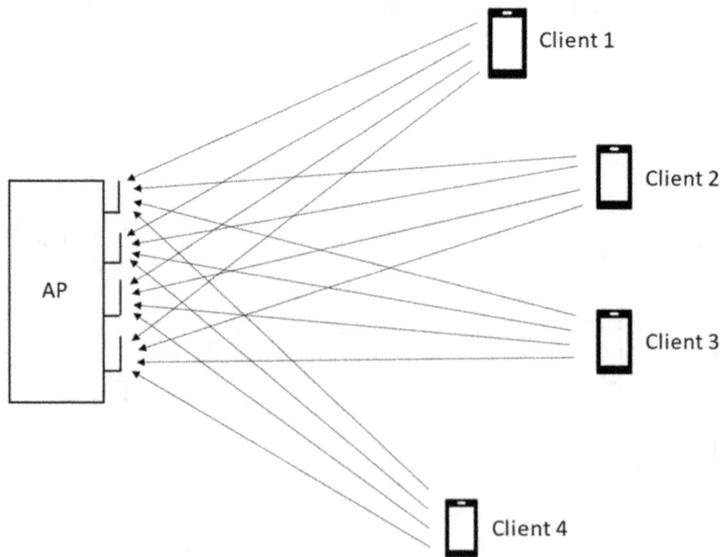

Fig. 9.8 An UL MU-MIMO system

derived channel information to the transmitter. The transmitter then computes the required channel matrix and uses this to focus the transmitted signal toward the receiver.

9.5.4   Modulation and Coding

9.5.4.1   Modulation

802.11ac (Wi-Fi 5) supported BPSK, QPSK, 16-QAM, 64-QAM, and 256-QAM modulation. 802.11ax (Wi-Fi 6) added 1024-QAM. As with 1024-QAM each symbol transmits 10 bits versus 8 bits transmitted by 256-QAM, 1024-QAM increases throughput by 25%. This improved throughput is achieved, however, at the expense of degraded BER versus E_b/N_0 (S/N) performance as can be seen in Fig. 4.11 where for a BER of 10^{-3} 1024-QAM requires an increase in E_b/N_0 of approximately 4.5 dB. Thus, the benefits of 1024-QAM can only be realized in short-range situations with good channel conditions. See Sect. 4.3 for details on the above-mentioned modulations.

Optional on 802.11ax is *Dual Carrier Modulation* (DCM). With DMC, the same signal is allocated to two subcarriers which are separated far apart in the frequency domain. DCM reduces the transmission rate by a factor of 2. However, it provides in effect frequency diversity increasing the received signal-to-noise ratio by about 3 dB, leading to increased robustness and an improvement in long distance coverage. The use of DCM is limited to BPSK, QPSK, and 16-QAM and is enabled only for long distance transmission. It is supported only in single and dual spatial stream modes.

9.5.4.2   Coding

In 802.11ax systems, packets are encoded with either Binary Convolution Code (BCC) coding (Sect. 5.3.5) or Low-Density Parity Check (LDPC) coding (Sect. 5.3.4). BCC is supported with some exceptions ([1], Sect. 27.1.1). LDPC is supported if certain conditions are met and is optional if a certain condition is met ([1], Sect. 27.1.1).

The basic BCC coder, shown in Fig. 9.9, is a rate 1/2 one of constraint length 7. The supported rates of 2/3, 3/4, and 5/6 are derived from it by employing puncturing (Sect. 5.3.6).

The LDPC coders defined are in fact Quasi-Cyclic LDPC (QC-LDCP) ones, with codeword block lengths of 648, 1296, and 1944, and lifting factors of 27, 54, and 81, respectively. For each codeword block length, the standard defines code rates of 1/2, 2/3, 3/4, and 5/6. The encoders are systematic, i.e., they encode an information block of length k by adding n-k parity bits to each information block to create a codeword of length n.

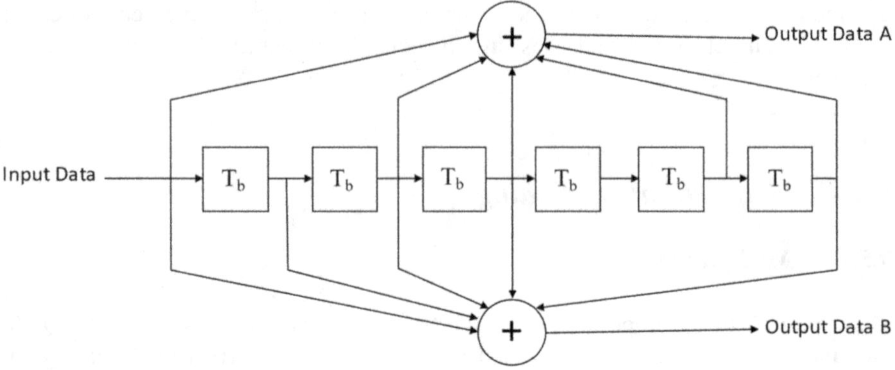

Fig. 9.9 Wi-Fi 6 convolution encoder

9.5.5   *Wi-Fi 6/6E Data Rates*

Very high data rates are one of the major features of Wi-Fi 6/6E. For systems such as Wi-Fi 6/6E, the data rate per spatial layer can be computed as follows:

$$\text{Data rate} = \frac{1}{T_{s+g}} \times Q \times N \times R \times S \tag{9.1}$$

where:

- T_{s+g} is the symbol time including the guard interval (and hence $1/T_{s+g}$ is the symbol rate).
- Q is the bits per symbol for the applied modulation scheme, being 10 for 1024-QAM, the highest order modulation supported.
- N is the number of data subcarriers.
- R is the coding rate.
- S is the number of Spatial Streams.

For 802.11ax:

- The minimum T_{s+g} and hence maximum symbol rate is for a guard interval of 0.8 µs. Here $T_{s+g} = (12.8 + 0.8) = 13.6$ µs.
- Maximum $Q = 10$ for 1024-QAM.
- Maximum number of data carriers $N = 1960$ for 160 MHz of channel bandwidth.
- Maximum code rate $R = 5/6$.
- Maximum number of spatial streams $S = 8$.

Applying the above data to Eq. (9.1) but for 1 spatial stream, i.e., $S = 1$, we get a maximum data rate of 1201 Mb/s. If we assume 8 spatial streams, then we get a maximum data rate of 9608 Mb/s or 9.6 Gb/s! Table 9.5 provides the 802.11ax approved modulation coding schemes (MCSs) and the maximum data rates for the supported channel bandwidths and one spatial stream.

Table 9.5 802.11ax maximum data rates (Mb/s, 0.8 µs GI) for 1 spatial stream

MCS	Modulation type	Code rate	20-MHz channel	40-MHz channel	80-MHz channel	160-MHz channel
0	BPSK	1/2	8.6	17.2	36.0	72.0
1	QPSK	1/2	17.2	34.4	72.1	144.2
2	QPSK	3/4	25.8	51.6	108.1	216.2
3	16-QAM	1/2	34.4	68.8	144.1	282.2
4	16-QAM	3/4	51.6	103.2	216.2	432.4
5	64-QAM	2/3	68.8	137.6	288.2	576.4
6	64-QAM	3/4	77.4	154.9	324.4	648.8
7	64-QAM	5/6	86.0	172.1	360.3	720.6
8	256-QAM	3/4	103.2	206.5	432.4	864.8
9	256-QAM	5/6	114.7	229.4	480.4	960.8
10	1024-QAM	3/4	129.0	258.1	540.4	1080.8
11	1024-QAM	5/6	143.4	286.8	600.5	1201.0

Mentioned in Sect. 9.5.4.1 above is the option of Dual Carrier Modulation (DCM). This is not given in Table 9.5 but is supported in MCS 0, 1, 3, and 4. When utilized, then in effect the code rate is halved and thus the associated data rate is also halved.

Although in theory any one user could communicate at the rates given in Table 9.5, in practice this rarely happens. The rates shown are better termed through-put rates. That is because the system typically serves multiple users and divides up capacity via OFDMA and MU-MIMO so that no one user accesses the full capacity. We note that in reality these maximum rates can be further diminished because the medium is a shared one and continuous access is not guaranteed.

9.5.6   Wi-Fi 6/6E PPDU Frame Formats

In the 802.11ax High Efficiency (HE) specification four different PPDU formats are defined:

- **Single user PPDU (HE_SU)**: Used when transmitting to a single user.
- **HE extended range PPDU (HE_EXT_SU)**: Used when transmitting to a single user but one further away from the AP such as in outdoor scenarios.
- **Multi-user PPDU (HE_MU)**: Used when transmitting to one or more users.
- **HE trigger-based PPDU (HE_TRIG)**: Sent in a single OFDMA and/or MU-MIMO uplink transmission in response to a Trigger frame sent down by an AP.

HE PPDUs consist of a preamble, a data portion, and a packet extension. The preamble consists of the Legacy Preamble for backward compatibility followed by the new HE Preamble. The Legacy Preamble is common to all four HE PPDU formats. Figure 9.10 shows the general frame format of HE PPDUs.

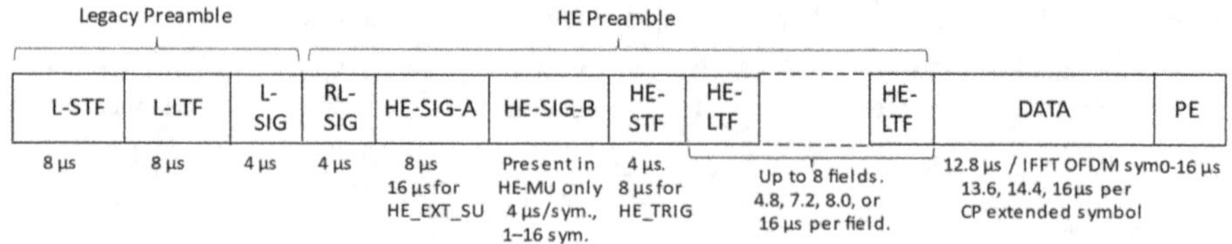

Fig. 9.10 HE PPDU general frame format

Following is a brief description of the various HE PPDU fields. Note that these descriptions involve OFDM symbols, the subcarriers in these symbols being modulated by binary sequences that convey the desired information.

- **L-STF** (Legacy short training field): 8 μs long, containing 2 IFFT OFDM symbols each 3.2 μs long plus a guard interval. This is the first piece of a PPDU packet that the receiver sees, is a true preamble, and it is used for automatic gain control (AGC), data stream time synchronization, and frequency offset correction only. It does not contain coded information.
- **L-LTF** (Legacy long training field): 8 μs long, containing 2 IFFT OFDM symbols each 3.2 μs long plus a guard interval. This is the second piece of a PPDU packet that the receiver sees, is also a true preamble, and it is used for channel estimation so as to decode the L-SIG field that follows and to improve the time synchronization and frequency offset correction. Like the L-STF, it does not contain coded information. All the following "preamble" fields, however, are not true preambles, but rather convey coded information necessary for the receiver to successfully decode that carried data.
- **L-SIG** (Legacy signal field): 4 μs long, containing 1 IFFT OFDM symbols plus a guard interval. It conveys the transmission time of the package.
- **RL-SIG** (Repeated L-SIG). It identifiers the frame to follow as a 802.11ax one.
- **HE-SIG-A** (HE signal A field): It is 8 μs long, containing 2 IFFT OFDM symbols each 3.2 μs long plus guard intervals for all formats except HE_EXT_SU. For HE_SU, HE_MU, and HE_TRIG PPDUs, the second OFDM symbol is BPSK modulated. For an HE_EXT_SU PPDU, the second OFDM symbol modulation is QBPSK (BPSK rotated by 90°). It contains all the transmission parameters for HE_SU and HE_EXT_SU users and common transmission parameters for HE_MU users. These parameters include whether the link is UL or DL, the BSS color (Sect. 9.6), bandwidth, modulation and coding, number of spatial streams, and the remaining time in the transmit opportunity. For HE_EXT_SU, the 2-symbol field is repeated once, increasing its length to 16 μs, in order to improve resilience in outdoor scenarios.
- **HE-SIG-B** (HE signal B field): Only present in HE_MU. Consists of up to 16 4 μs guard interval extended symbols, its length being determined by the number of clients it is addressing. It begins with a Common field that contains RU allocations, the number of users and other information common to all transmissions. The Common field is followed by a number of User Specific fields that contains each client's key information such as client identification, modulation and coding scheme, number of spatial streams, etc.

- **HE-STF** (HE short training field): Consists of one 4-µs guard interval extended symbol for all formats except HE_TRIG when it is two such symbols. It is applicable to MIMO operations only and aids the receiver in synchronizing with the incoming frame in time and frequency before decoding the rest of the packet.
- **HE-LTF** (HE long training field): There can be up to eight such fields, with one symbol each of total length of either 4.8 (3.2 + 1.6), 7.2 (6.4 + 0.8), 8.0 (6.4 + 1.6), or 16 (12.8 + 3.2) µs. It is applicable to MIMO operations only and is responsible for beamforming and spatial diversity.
- **Data** (Data field): MPUD from the MAC layer above.
- **PE** (Packet extension field): 0, 4, 8, 12, or 16 µs long. Used to provide the receiver with extra time to process the frames contents before responding with its own generated frame.

As noted above, the detection of RL-SIG identifiers the frame to follow as an 802.11ax one. This is only the first step, however, in frame identification. Recall that there are four different PPDU formats for 802.11ax. To determine which of the four possible frames is being detected, the receiver executes the following procedure:

- A computation involving the transmission time of the package that determines if the format is either (a) HE_SU or HE_TRIG or (b) HE_MU or HE_EXT_SU.
- If determined to be HE_SU or HE_TRIG, then the first data bit in the HE-SIG-A field determines which of these two formats is present.
- If determined to be HE_MU or HE_EXT_SU, then if the second OFDM symbol of the HE-SIG-A field is BPSK modulated, the format is determined to be HE_MU. However, if the second OFDM symbol of the HE-SIG-A field is QBPSK modulated, the format is determined to be HE_EXT_SU.

9.5.7   *Spatial Reuse (SR) and BSS Coloring*

The normal way in which 802.11 works is that, before commencing communication, every station must listen and ascertain if the channel is idle. If so, the communication can commence. However, if determined to be busy, then communication must be delayed until the channel is determined to be idle. Because all Wi-Fi communication takes place over unlicensed channels the probability of a busy channel is real. If we take the 2.4-GHz band, for example, there are only 3 non-overlapping 22-MHz channels. Thus, in a high user density environment, where the need for more than three APs are required, adjacent APs using the same channel is likely inevitable.

To improve the efficient use of available spectrum and hence system level performance in dense utilization scenarios, 802.11ax introduced a *Spatial Reuse* (SR) technique. Here stations in a Basic Service Set (BSS) can identify if a signal occupying the spectrum is from its service set or from an Overlapping Basic Service Set (OBSS) and take a decision as to whether to defer to this signal or transmit in the channel despite its presence. Figure 9.11 shows a BSS and an OBSS, both using the same 20-MHz channel. Here BSS User 1 is likely the at times to detect interfering signals from AP2 and OBSS User 1.

802.11ax determines if a signal is from the BSS or the OBSS by using *BSS coloring*. BSS color is not a color as such but is a six-bit sub field in the HE SIG-A field of the PPDU preamble that identifies the specific BSS.

When a station that is actively monitoring the medium detects an 802.11ax frame, it checks the color encoded in that frame. If the color is the same as the one that its associated AP has signified, then the station concludes that the frame is an *intra-BSS frame*. However, if the detected frame has a different BSS color than its own, then the station concludes that the frame is an *inter-BSS frame* from an OBSS.

When a station detects an Intra-BSS frame, it concludes that the channel is busy and does not attempt at that time to commence communication. However, if the detected frame is an inter-BSS frame, then it must decide if it can ignore the presence of this signal and commence communication. For a 20-MHz channel, a simplified version of how it makes this decision as follows:

- Station detects a signal.
- Station determines if the signal is above or below a threshold called the *Clear Channel Assessment* (CCA) threshold. For a 20-MHz channel, this threshold is normally defined as −82 dB.
- If the signal level is below the CCA threshold, the channel is deemed idle and transmission can commence.
- If the signal is above the CCA threshold, the station checks if it is an 802.11ax type signal and thus can demodulate traffic.
- If no, then if the level is below the so-called energy threshold, which for a 20-MHz channel is −62 dBm, the channel is deemed idle.
- If yes, it decodes the frame header to determine the color of the frame.

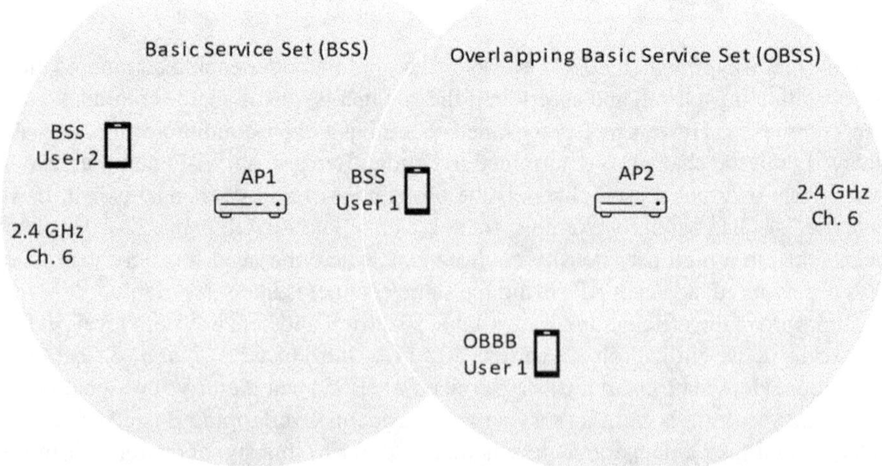

Fig. 9.11 A BSS and OBSS

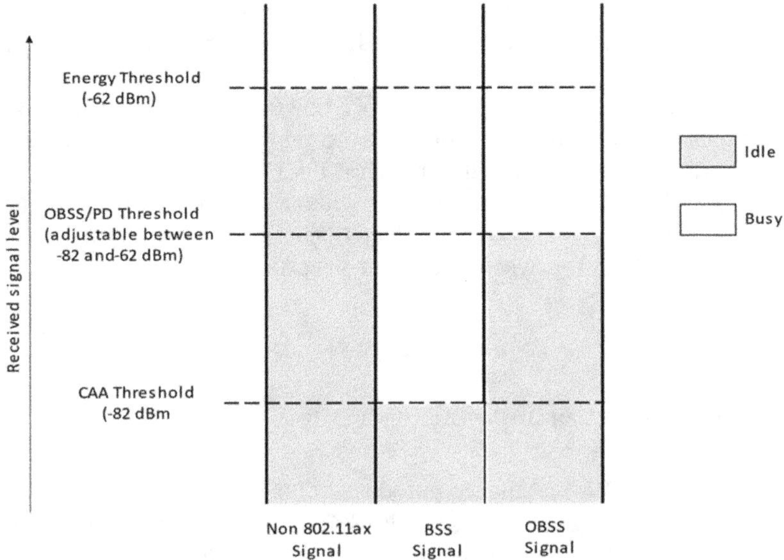

Fig. 9.12 Busy and idle states relative to received signal level

- If the color is the same as its associated AP, it means that the frame is an intra-BSS one and the channel is deemed busy.
- If the color is not the same as its associated AP, then the frame is deemed an inter-BSS one.
- The station then checks if the signal strength of the frame is above the energy threshold which is −62 dBm. If it is, then the channel is deemed busy.
- If it is lower than −62 dBm, then the station will adjust the CCA threshold upwards to a higher level called the OBSS/PD level which is above the received signal level and in so doing allow for transmission. This increase in threshold level is accompanied, however, by a decrease in transmit power. The higher the OBSS/PD threshold, the lower the transmit power, this being done in order to minimize interference in the OBSS.

Figure 9.12 shows busy and idle states relative to received signal level for BSS colored, OBSS colored, and non 802.11ax signals.

Note that a doubling of bandwidth results in a +3 dB increase in both CCA threshold and energy threshold. Thus, for a 40-MHz channel, the CCA threshold increases to −79 dBm and the energy threshold increases to −59 dBm.

9.5.8 Target Wake Time (TWT)

802.11ax supported Target Wake Time (TWT) allows an AP to manage the activity in its BSS so as to minimize contention between STAs and reduce the amount of time that a STA using a power management system mode is required to be awake.

An AP achieves this by scheduling STAs to operate at non-overlapping times and/or frequencies as well as concentrating AP-STA frame exchanges in predetermined service periods.

An AP can coordinate the use of the TWT function with participating STAs in order to define a specific time or set of times for individual STAs to access the RF medium. Further, via the so termed "Broadcast TWT" operation, an AP can provide schedules and deliver TWT times to STAs without establishing individual agreements with them. TWT increases STA sleep time. This is of particular importance to battery operated devices where increased sleep time translates to extended battery life.

9.6   Wi-Fi 7 Key Technologies

In January 2024, the Wi-Fi Alliance introduced "Wi-Fi CERTIFIED 7." This certification is ahead of the IEEE release of its 802.11be Extremely High Throughput (EHT) standard which is expected before the end of 2024. Nonetheless, devices that meet the Wi-Fi CERTIFIED 7 tests will also meet the essential technical specifications of 802.11be. The new technologies in Wi-Fi 7 aim to provide increased data rates along with improved reliability and reduced latency. These new capabilities will be key to supporting new applications and services such as 4K/8K video streaming, augmented reality (AR), and virtual reality (VR). Informative material on Wi-Fi 7 can be found in [5, 6].

9.6.1   *320-MHz Channel Bandwidth*

Wi-Fi 7 operates in the same supported bands as Wi-Fi 6E, namely 2.4, 5, and 6 GHz. With Wi-Fi 6E, the maximum channel bandwidth was 160 MHz. With Wi-Fi 7 this maximum is doubled to 320 MHz, this change alone allowing a doubling of the maximum data rate relative to Wi-Fi 6/6E.

In Wi-Fi 7, the supported RU sizes corresponding to the entire 20-, 40-, 80-, 160-, and 320-MHz bandwidths are 242, 484, 996, 2x996, and 4x996 tones, respectively.

9.6.2   *4096-QAM Modulation*

Wi-Fi 6/6E supported a maximum QAM constellation of 1024 points. Wi-Fi 7 increases this to 4096 points. As given in Table 4.1, this increases the maximum spectral efficiency to 12 bits/s/Hz from 10 bits/s/Hz resulting in an increase in capacity of 20% for a given bandwidth and the same coding rate. We note also from

Table 4.1 that this increase in spectral efficiency requires a 5 dB increase in E_b/N_0 relative to 1024-QAM to maintain a BER of 10^{-3}.

9.6.3   16 × 16 MIMO

With Wi-Fi 6, the maximum number of AP antennas supported is 8. An AP equipped with this maximum can simultaneously support up to eight spatial streams and hence serve, via MU-MIMO (Sect. 7.3.4), up to eight users in both the downlink (DL) and uplink (UL). With Wi-Fi 7 the maximum number of AP antennas supported has been increased to 16 and hence the maximum number of spatial streams supported up to 16. This not only increases the maximum number of users that the AP can serve to 16 but also increases the maximum throughput by a factor of 2.

9.6.4   Wi-Fi 7 Data Rates

As indicated in Sect. 9.5.5, Wi-Fi data rate is a function of:

- T_{s+g}, the OFDM symbol time, including the guard interval (and hence $1/T_{s+g}$ is the symbol rate).
- Q, the bits per symbol for the applied modulation scheme, being 10 for 1024-QAM and 12 for 4096-QAM.
- N, the number of data subcarriers.
- R, the is the coding rate.
- S, the number of spatial streams.

  With Wi-Fi 7:

- OFDM symbol length and guard interval options are the same as for Wi-Fi 6. Hence, the minimum T_{s+g} and hence maximum symbol rate is for a guard interval of 0.8 µs. Here $T_{s+g} = (12.8 + 0.8) = 13.6$ µs.
- $Q = 12$.
- Maximum number of data carriers N is 3920 for a 320-MHz channel bandwidth.
- Coding rates supported for 4096-QAM are the same as those supported for 1024-QAM, namely 3/4 and 5/6. Thus, maximum coding rate for Wi-Fi 7 is 5/6.
- Maximum number of spatial streams is 16.

Applying the above data to Eq. (9.1) we get a maximum data rate for Wi-Fi 7 of 46 Gb/s. This is not surprising, given that for Wi-Fi 6, the equivalent rate is 9.6 Gb/s, and with Wi-Fi 7, the rate is increased by 2 because of increased channel bandwidth, by 1.2 because of higher modulation index, and by 2 because of the doubling of the maximum number of spatial streams, these increases collectively leading to an increase by a factor of $2 \times 1.2 \times 2 = 4.8$, and $9.6 \times 4.8 = 46$.

Table 9.6 802.11be maximum data rates (Mb/s, 0.8 μs GI) for 4096-QAM and 1 spatial stream

MCS	Modulation type	Code rate	20-MHz channel	40-MHz channel	80-MHz channel	160-MHz channel	320-MHz channel
12	4096-QAM	3/4	154.8	309.7	648.5	1297.0	2594.0
13	4096-QAM	5/6	172.1	344.2	720.6	1441.2	2882.4

Table 9.6 provides, for 802.11be and 4096-QAM, the approved modulation and coding schemes (MCSs) and associated the maximum data rates for the supported channel bandwidths and one spatial stream.

9.6.5 Multi-link Operation (MLO)

Multi-Link Operation (MLO) is arguably the most important new feature of Wi-Fi 7. It provides the ability, depending on how applied, to either increase capacity, enhance reliability, or reduce latency. A number of pre-Wi-Fi 7 versions allow operation in multiple bands. For example, Wi-Fi 6E, like Wi-Fi 7, allows operation in three bands, namely the 2.4-, 5-, 6-GHz bands. With all these pre-Wi-Fi 7 versions, however, operation at any one time is limited to one band, only switching to another if conditions change. This restriction leaves the other supported band/s unused. With MLO, a Wi-Fi 7 enabled AP and non-AP device can communicate over multiple channels and frequency bands simultaneously. See Fig. 9.13.

Multi-channel simultaneous communication capability is supported in three different operating modes, each providing a specific advantage. These modes are link aggregation, link redundancy, and link selection.

With link aggregation, data is split and sent over different radios operating on different bands. As a result, the amount of data being sent at the same time is increased, i.e., the overall throughput is increased. Note, however, that with operation in this mode, neither reliability nor latency is improved.

With link redundancy, reliability is increased having the device send the same data over multiple bands using different radios. At the receiving end, the data frames are compared and only that which is error free is forwarded. By this comparison, the receiving unit has an improved chance of sending an ACK message back to the transmitting device after receiving each frame, allowing a new one to be transmitted. In summary, by sending the same data over multiple channels, the chances of receiving the correct data are improved, hence improved reliability.

Link selection is a slight variation of link redundancy. Here, on a regular basis, the transmitting device evaluates the performance of the available channels on the different bands and sends the data on the channel with the best performance, thus enhancing the reliability of transmission. By only using one channel most of the time, other channels are freed up, allowing more devices to operate simultaneously with a single AP.

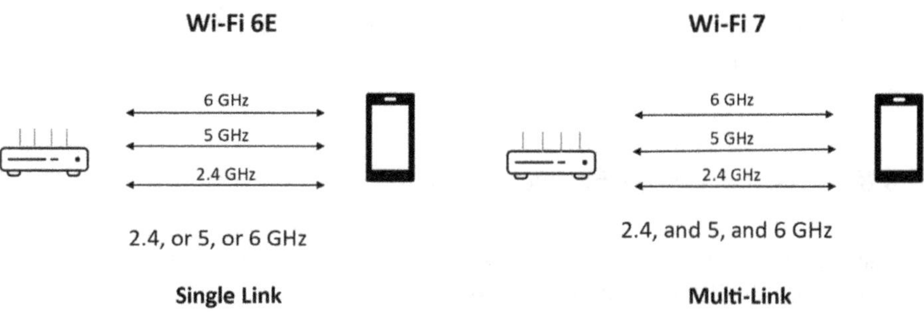

Fig. 9.13 Single link and multi-link operation

9.6.6 Multi-RU

In Wi-Fi 6/6E each user is only assigned one RU (Sect. 9.5.1.2) for transmitting and receiving frames. This significantly limits the flexibility of the spectrum resource scheduling. To solve this problem and further improve the spectral efficiency, Wi-Fi 7 allows the allocation of multiple RUs to a single user and allows the combination of RUs of different sizes. However, to achieve a trade-off between combination complexity and spectral efficiency, only defined combinations of RUs are allowed. Small-size RUs (<20 MHz, i.e., 26, 52, and 106-tone) can only be combined with small-size RUs, and large-size RUs ($\geq$20 MHz, i.e., $\geq$242-tone) can only be combined with large-size RUs. Combinations of small-size and large-size RUs are permitted.

9.6.7 Preamble Puncturing

Preamble puncturing capability is optional in Wi-Fi 6/6E but is not commonly used in most Wi-Fi 6/6E products. It is mandatory in Wi-Fi 7. It allows an access point (AP) to remove or "puncture" a segment of a channel that's affected by interference from another AP, allowing the rest of the channel to be available for transmission. The size of the punctured segment is specified as a multiple of 20 MHz.

Figure 9.14 illustrates a puncturing example where an AP is operating in an 80-MHz channel and where there is interference present in a 20-MHz segment. With no puncturing, the AP can only use the first 40 MHz of the channel. However, with puncturing, it is now able to use the first 40 MHz as well as the last 20 MHz.

Note that with Wi-Fi 6/6E puncturing, only multi-user transmission is allowed. Thus, in our example, the first two 20-MHz channels could be used for transmission with User 1 say, while the last 20-MHz channel would have to be used for transmission with User 2 say. This is because multi-RU operation, i.e., multi-RUs to a single user, is not supported by Wi-Fi 6/6E. With Wi-Fi 7, on the other hand, where multi-RU operation is supported, the first two and the last 20-MHz channels can all be used for transmission with a single user.

9.6.8   Wi-Fi 7 PPDU Frame Formats

As described in Sect. 9.5.6, four PPDU formats are defined for Wi-Fi 6/6E. For Wi-Fi 7 only two PPDU formats are defined:

– Multi-user PHY protocol data unit (EHT-MU PPDU).
– Trigger based PHY protocol data unit (EHT-TB PPDU).

The EHT MU PPDU is sent to a single or to multiple users, whereas with Wi-Fi 6/6E different PPDUs are sent to a single versus multiple users. Figure 9.15 shows the EHT MU PPDU format. Relative to the Wi-Fi 6/6E frame formats, the legacy and RL-SIG fields are unchanged. Following the RL-SIG field, the HE-SIG-A and HE-SIG-B fields are replaced with the new Universal SIG (U-SIG) and EHT-SIG fields.

The U-SIG field contains the parameters necessary to interpret EHT PPDUs. It consists of a version independent field and a version dependent field. The version independent field consists of information such as the PPDU type, MCS, bandwidth, and BSS color. The version dependent field contains information similar to that

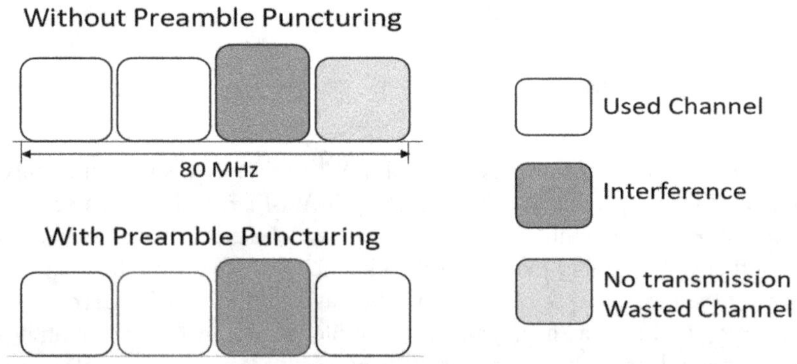

Fig. 9.14 Example of preamble puncturing

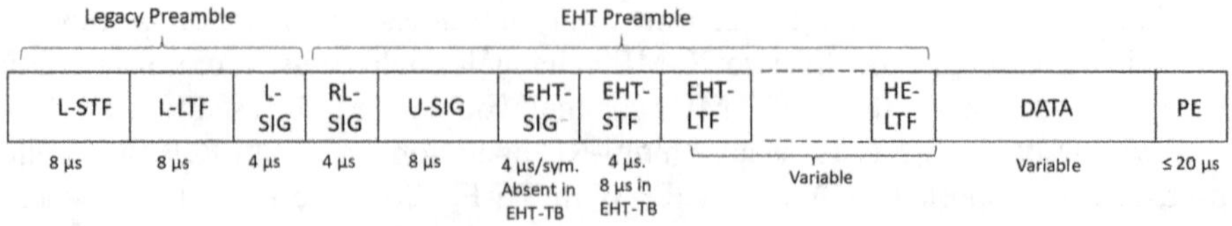

Fig. 9.15 EHT MU PPDU format

included in HE-SIG-A, except for information included in the version independent field. It also contains new information such as guard interval duration, etc.

The EHT-SIG field consists of a common field and a user specific field. The common field contains common information about the PPDU such as RU allocation, coding, MSC, number of spatial streams, guard interval duration, etc. The user specific field contains dedicated information for individual users.

Following the ETH-SIG field, the HE-STF and HE-LTF fields are replaced with the EHT-STF and EHT-LTF fields whose functions are similar to those of the HE equivalents.

Similar to the Wi-Fi 6/6E HE Trigger Based PPDU, Wi-Fi defines an ETH Trigger Based PPDU (EFT TB PPDU). Its format is similar to the EHT MU PPDU, with the exception that it does not contain the EHT-SIG field and the EHT-STF field is two times longer than in the EHT MU PPDU. The AP uses this frame to allocate resources to STAs and solicit a response from one or more STAs. The STAs use it to respond to the EHT-TB received from the AP.

9.6.9 Multi-AP Coordination

As the name implies, Multi-AP coordination involved coordination between nearby APs. The goal is to optimize channel selection and vary the loads between APs so as to achieve efficient utilization and a balanced allocation or radio resources. Multi-AP transmission modes include:

- Coordinated Spatial Reuse (CSR): Here APs cooperatively control their transmit power in a coordinated fashion so as to have adequate SNR at all stations and minimize interference in order to maximize the network-wide throughput. Key to CSR is coordinated scheduling. Such scheduling requires coordinated scheduling measurement which allows APs to get information about other co-channel APs in a timely fashion.
- Coordinated OFDMA (C-OFDMA): With this approach, APs coordinate their schedules in frequency and time with each other so as to share OFDMA resources for all stations, avoiding RU conflicts and improving spectrum utilization.
- Coordinated Beamforming (CBF): The principle behind this approach is that an AP, while generating beams for transmission to its associated stations, also endeavors to null its interference to nearby non-associated stations.
- Joint Transmission (JTX): For stations that are able to receive from multiple APs, this approach permits multiple APs to serve the same station by employing virtual MU-MIMO. It enables fast connection of stations with an optimal AP and improves the speed of reconnection as users move around.
- Coordinated TDMA: Here the APs take turn in transmitting to their associated stations on the same resource units.

The Wi-Fi 7 drafters considered all of the above stated multi-AP transmission modes. However, only a low complexity version of coordinated spatial reuse (CSR) became part of the Wi-Fi 7 (802.11be) standard. A proper suite on Multi-AP coordination modes is now expected as part of Wi-Fi 8.

9.7 Comparison Between Wi-Fi 7, Wi-Fi 6E, and Wi-Fi 5

Above the key features of 802.11ax (Wi-Fi 6E) and 802.11be (Wi-Fi 7) have been presented. With an understanding of these features, a comparison with key features of 802.11 ac (Wi-Fi 5), the predecessor release, is in order. Such a comparison is presented in Table 9.7.

Table 9.7 A comparison of key features of 802.11be, 802.11ax, and 802.11ac

Feature	Wi-Fi 5 (802.11 ac)	Wi-Fi 6E (802.11ax)	Wi-Fi 7 (802.11be)
Release date	2013	2020	2024
Frequency bands	5 GHz	2.4, 5, 6 GHz	2.4, 5, 6 GHz
Channels	20, 40, 80, 80 + 80, 160 MHz	20, 40, 80, 80 + 80, 160 MHz	20, 40, 80, 80 + 80, 160, 160 + 80, 240, 160 + 160, 320
Multi access scheme	OFDM	OFDMA	OFDMA
Subcarrier spacing	312.5 kHz	78.125 kHz	78.125 kHz
OFDM symbol length	3.2 µs	12.8 µs	12.8 µs
OFDM guard interval	0.8, 0.4 µs	0.8, 1.6, 3.2 µs	0.8, 1.6, 3.2 µs
Highest order modulation	256-QAM	1024-QAM	4096-QAM
Dual carrier modulation	Not available	Supported	Supported
BCC coding	Mandatory	Mandatory	Mandatory
LDPC coding	Optional support	Optional support	Mandatory
Max. spatial streams	8	8	16
MU-MIMO	DL only	DL and UL	DL and UL
Max. throughput (Gb/s)	6.93 Gb/s (8 SS)	9.61 Gb/s (8 SS)	46.1 (16 SS)
Target wait time (TWT)	Not available	Supported	Supported
Multi-link operation	Not supported	Not supported	Supported
Multi-RUs per station	Not supported	Not supported	Supported

9.8 Summary

Wi-Fi 6/6E (802.11ax High Efficiency), the release following Wi-Fi 5 (802.11ac Very High Throughput), was designed to facilitate high-density wireless access and high-capacity wireless services, in locations such as high-density sports arenas, indoor high-density offices, and large-scale outdoor public venues. Wi-Fi 7 (802.11ax Extremely High Throughput) the release following Wi-Fi 6E aims to provide increased data rates along with improved reliability and reduced latency.

In this chapter, we first examined, for Wi-Fi in general, Internet access network architecture and protocol architecture. This was followed by a study of the RF spectrum supported by both Wi-Fi 6/6E and Wi-Fi 7. Next, for Wi-Fi 6/6E, we reviewed the physical layer and key technologies employed therein, and some new procedures for improving medium utilization and battery efficiency. We then examined the key features of Wi-Fi 7, including Multi-Link Operation (MLO), arguably its most important new feature. Finally, a comparison between Wi-Fi 7, Wi-Fi 6E, and Wi-Fi 5 was presented.

References

1. IEEE Std. 802.11ax (2021) Part 11: wireless LAN medium access control (MAC) and physical layer (PHY) specifications, amendment 1: enhancements for high-efficiency WLAN. IEEE, New York
2. IEEE Std. 802.11 (2020) Part 11: wireless LAN medium access control (MAC) and physical layer (PHY) specifications. IEEE, New York
3. Deng D-J et al (2017) IEEE 802.11ax: highly efficient WLANs for intelligent information infrastructure. IEEE Commun Mag 55(12):52–59
4. Khorov E et al (2019) A tutorial on IEEE 802.11ax high efficiency WLANs. IEEE Commun Surv Tutor 21(1):197
5. Cailain D et al (2020) IEEE 802.11be Wi-Fi 7: new challenges and opportunities. IEEE Commun Surv Tutor 22(4):2136
6. Ward L, Kopp J (2022) IEEE 802.11be technology introduction. Rohde & Schwarz, Munich

Chapter 10
Bluetooth 5/6 Overview

10.1 Introduction

As indicated in Sect. 1.4, Bluetooth is a short-range wireless technology used for the exchange of data between devices, fixed or mobile, over short distances (a few meters up to about 240 m). It operates in the 2.4-GHz unlicensed band and is primarily used as an alternative to wired connections to link, for example, smartphones and music players to headphones and speakers.

In this chapter, emphasis is placed on the latest two versions of Bluetooth, namely version 5 and 6, both of which fall under the nomenclature of *Bluetooth Low Energy* or BLE. BLE was first introduced in Version 4.0 of the Bluetooth Core Specification. Since then, BLE has been expanded in its capabilities via Versions 4.1, 4.2, 5.0, 5.1, 5.2, 5.3, 5.4, and now 6.0. A key feature of BLE is that it is highly efficient in its use of power, hence its nomenclature. This allows a new class of devices that are powered by small coin-sized batteries and that can operate for days, weeks, or even longer without a battery change. In addition, BLE supports not only the point-to-point topology, but also mesh and broadcasting topologies. Earlier versions, known collectively as *Bluetooth Classic*, only supported the point-point topology. BLE in the broadcast mode permits one transmitting device to send data to an unlimited number of receiving devices. In the sections below, we examine, for BLE, the protocol architecture, the RF spectrum supported, the physical layer and link layer (LL) and the key technologies employed therein, and finally a comparison between BLE and Bluetooth Classic is presented.

Most of the material presented in this chapter is sourced from Bluetooth Core Specifications Versions 5.4 and 6.0 [1, 2], Bluetooth Core Specifications, Version 5.4, Technical Overview [3], Bluetooth Core Specification, Version 6.0, Feature Overview [4], and The Bluetooth Low Energy Primer [5]. For those readers seeking somewhat more detail than presented here [5] is highly recommended. For those desiring an in-depth and detailed description in any given area then [1–4] is recommended.

D. H. Morais, *5G/5G-Advanced, Wi-Fi 6/7, and Bluetooth 5/6*, https://doi.org/10.1007/978-3-031-82830-0_10

10.2   Protocol Architecture

As indicated in Sect. 2.3, instead of applying the TCP/IP model, Bluetooth LE has its own independent protocol stack. However, there are many similarities such as the use of packets and frames and a certain equivalence between the Bluetooth LE layers and those of TCP/IP. The functionality of this stack is divided into three main building blocks: the *Application Layer*, the *Host*, and the *Controller*, and a standard logical interface between the Host and the controller that defines the way in which the Host and Controller may communicate. The Host and the Controller behave as separate logical entities which may be realized in physically separate components. Figure 10.1 shows the Bluetooth LE protocol stack. Starting from the top we have:

– **The Application (APP) Layer**: This layer is the direct user interface that defines profiles that allow interoperability between various user applications. Interoperability ensures that an application can effectively communicate with other applications on different devices.

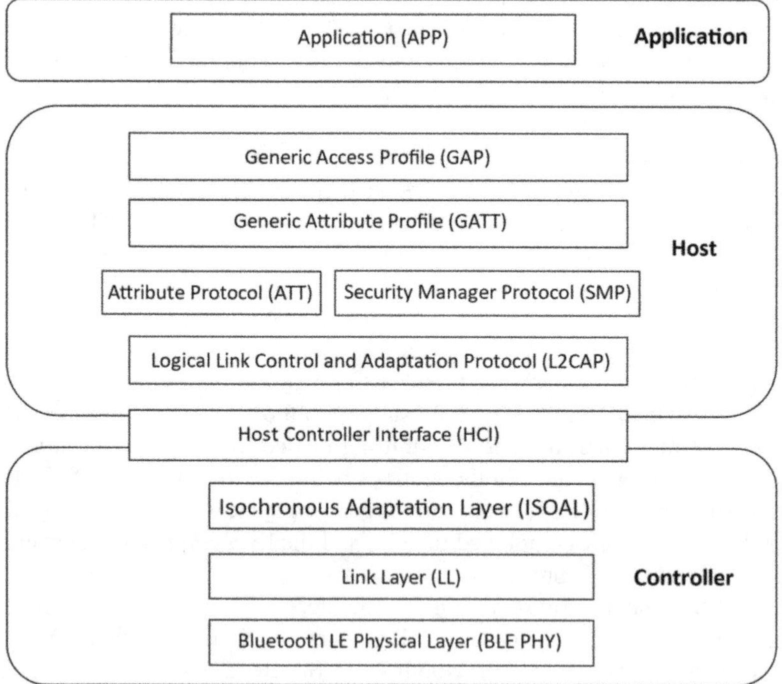

Fig. 10.1   Bluetooth LE protocol stack

- **The Generic Access Profile (GAP)**: This protocol, which resides within the Host, is the part of the architecture that defines security levels and how BLE devices communicate with each other when in a non-connected state. It interfaces directly with the APP layer above.
- **The Attribute (ATT) Protocol**: This protocol, which resides within the Host, provides the means to transmit data between BLE devices. It relies on a BLE connection and provides procedures to read, write, and transmit "attribute" values over the connection. Application data is identified via attributes and GATT defines and creates appropriate attributes for a given application. The ATT creates outgoing and processes incoming packets that define the action in the GATT layer.
- **The Generic Attribute (GATT) Profile**: This profile, which resides within the Host, is built on top of the ATT and creates a common framework for the data that's stored and transported by the ATT. It defines how attributes are formatted, packaged, and sent across connected devices according to a set of rules. In addition to actual application data, it provides information about the attributes such as how they are accessed and what level of security, if any, is needed.
- **Security Manager Protocol (SMP)**: This protocol, which resides within the Host, applies security algorithms to encrypt/decrypt data packets and defines the methods and protocol for pairing of devices.
- **Logical Link Control and Adaptation Protocol (L2CAP)**: This protocol resides at the lowest level within the Host. It is responsible Quality of Service (QoS), routing, and encapsulation of data from higher layers into standard BLE packet format for transmission and de-encapsulating data from standard BLE packets during reception. Also, it serves as a protocol multiplexer, ensuring that protocols are serviced by the correct Host component.
- **Host Controller Interface (HCI)**: The HCI defines a set of commands and events for transmission downwards and reception upwards of packet data. For downwards transmission, it translates raw data into packets and sends these packets to the Controller. For upwards transmission, it translates packets into raw data and forward this data to the Host.
- **Isochronous Adaptation Layer (ISOAL)**: This layer permits different frame durations to be employed by devices using isochronous communication. It provides segmentation and reassembly, and fragmentation and recombination of packets to and from a higher layer.
- **Link Layer (LL)**: The LL, which is a part of the Controller, functions similar to the MAC layer in 5G NR and Wi-Fi 6. It interfaces directly with the physical layer and controls the state of the device (Sect. 10.4.1). It is also responsible for scheduling, the RF channel to be on, packet formats, error checking, etc.
- **Bluetooth LE Physical (BLE PHY)**: This layer is a part of the Controller and resides at the bottom of the BLE protocol stack. It is responsible for the actual transmission and reception of information over the air utilizing the 2.4-GHz unlicensed band. It applies associated technologies such as modulation, transmitter, and receiver characteristics, and frequency channel selection.

10.3   The Physical Layer

In this section, we explore the key features of the BLE version 5 physical layer so as to understand how BLE communicates physically across the assigned RF spectrum. We start by describing the supported frequency band and how it is channelized for BLE transmission then move on to specific hardware features.

10.3.1   Supported Frequency Band and Channelization

BLE operates in the 2.4-GHz unlicensed band. This band extends from 2400 to 2485.5 MHz and for BLE purposes is divided into 40 adjacent channels, each 2 MHz wide, as opposed to Bluetooth Classic, where there are 79 channels, each 1 MHz wide. These channels are used for carrying application data and *advertising*. Advertising allows devices to broadcast information that defines their intentions. In the Core Specification Version 4, *legacy advertising* employing three channels is used for advertising. However, in Core Specification Version 5.0, *extended advertising* is introduced which uses all 40 BLE channels, providing improved efficiency and less susceptibility to error due to packet collisions. Table 10.1 shows channel numbering, Data Channel Index, and Advertising Channel Index for Legacy Advertising. The Legacy advertising channels are RF Channel numbers 0, 12, and 39, and are referred to as the *primary advertising channels*. These channels are positioned so that they are not disturbed by the nonoverlapping Wi-Fi channels 1, 6, and 11 in this band (Sect. 9.4.1). The remaining 37 channels, when used for carrying data, are referred to as *general purpose channels* and when used for advertising are referred to as *secondary advertising channels*.

Table 10.1   RF channel mapping to data channel and advertising index for legacy advertising

RF channel #	RF center freq. (MHz)	Channel type	Data channel index	Advertising channel index
0	2402	Advertising		37
1	2404	Data	0	
2	2406	Data	1	
....	....	Data	....	
11	2424	Data	10	
12	2426	Advertising		38
13	2428	Data	11	
14	2430	Data	12	
....	....	Data	....	
38	2478	Data	36	
39	2480	Advertising		39

10.3.2   Modulation Scheme

The BLE PHY layer, in order to RF modulate, prior to transmission, digital data from the logical link (LL) layer immediately above and to demodulate received radio signals, uses Gaussian Frequency Shift Keying (Sect. 4.4). Below (Sect. 10.3.3) four physical variants of BLE, present in all iterations of version 5, are described.

Three of these variants, namely LE 1M, LE Coded $S = 2$, and LE Coded $S = 3$, have a symbol rate of 1 Mb/s. For these variants, the bandwidth-bit period product $BT = 0.5$, and the modulation index, m, is specified to be between 0.45 and 0.55. For GFSK, the bit period and the symbol period are equal as GFSK is a binary modulation scheme. Thus, for a 1 Msym/s symbol rate, the bit period is 10^{-6} s and $B = 0.5/10^{-6} = 0.5$ MHz. From Eq. (4.13), we have frequency deviation $\Delta f = m \cdot f_m$, where here $f_m = B$. Applying $m = 0.5$, the average modulation index specified, we get a nominal frequency deviation of +/−250 kHz. Further specified is a frequency deviation of at least +/−185 kHz.

The fourth variant is LE 2M which has a symbol rate of 2 Msym/s, and like the other variants, $BT = 0.5$ and the modulation index, m, between 0.45 and 0.55. Using the same analysis as above we determine that the nominal frequency deviation is +/−500 kHz, the specified minimum frequency deviation here being +/−370 kHz.

10.3.3   Physical (PHY) Variants

The PHY variants are summarized as follows:

- **LE 1M**: Here the symbol rate is 1 Msym/s, the nominal frequency deviation +/−250 kHz, the minimum frequency deviation +/−185 kHz/s, and no coding is applied. It must be supported by all devices.
- **LE 2M**: Here the symbol rate is 2 Msym/s, the nominal frequency deviation +/−500 kHz, the minimum frequency deviation +/−370 kHz/s and no coding is applied. Since the symbol rate is twice that of LE 1M, the time to transmit a given number of packets is cut in half resulting in improved spectral efficiency. Support is optional.
- **LE Coded $S = 2$**: Here the symbol rate is 1 Msym/s, the nominal frequency deviation +/−250 kHz, and the minimum frequency deviation +/−185 kHz/s. However, here FEC coding is applied. The coder used is a rate 1/2 Binary Convolutional Encoder (Sect. 5.3.5) with a constraint length $k = 4$. Its structure is illustrated in Fig. 10.2 and is followed by a *pattern mapper*, but for this variant the output of the mapper is the same as the input, i.e., an input of 0 results in an output of 0, an input of 1 results in an output 1. Thus, a single bit into the encoder results in 2 bits out of the pattern mapper. As a result, the input data rate (Protocol Data Rate) from the layer above must be reduced to 500 kb/s to keep the symbol

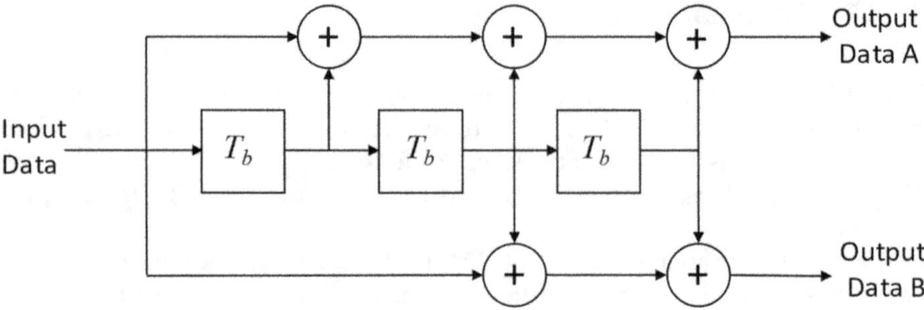

Fig. 10.2 BLE binary convolution encoder

Table 10.2 Comparison of the BLE PHY variants

	LE 1M	LE 2M	LE coded $S = 2$	LE coded $S = 8$
Modulation	GFSK	GFSK	GSFK	GSFK
Symbol rate	1 Mb/s	2 Mb/s	1 Mb/s	1 Mb/s
Protocol data rate	1 Mb/s	2 Mb/s	500 kb/s	125 kb/s
Approx. maximum[a] Application data rate	800 kb/s	1420 kb/s	380 kb/s	110 kb/s
Nominal frequency deviation	+/−250 kHz	+/−500 kHz	+/−250 kHz	+/−250 kHz
Error detection	CRC	CRC	CRC	CRC
Error correction	None	None	FEC	FEC
Approx. range multiplier	1	0.8	2	4
Requirement	Mandatory	Optional	Optional	Optional

[a] Computed in Sect. 10.4.4

rate at 1 Msym/s. Though the useful data rate has been halved, range is approximately doubled. Support is optional.

- **LE Coded $S = 8$**: This is similar to LE Coded $S = 2$ with the exception that the pattern mapper behaves differently. Here, an input of 0 results in an output of 0011 and an input of 1 results in an output of 1100. The net result is that a single bit into to the encoder results in eight bits out of the pattern mapper for an effective coding rate of 1/8. As a result, the Protocol Data Rate must be reduced to 125 kb/s to keep the symbol rate at 1 Msym/s. Though the useful data rate has been reduced by a factor of 8, range is approximately quadrupled. Support is optional.

The initial state of the encoder is set to all 0s. An input sequence of three consecutive 0s known as the *termination sequence* brings the encoder back to its original state.

For all four variants a cyclic redundancy check (CRC) (Sect. 5.2) is added for error detection. If the CRC check fails in the receiver and acknowledgment of this to the transmitter may cause the transmitter to resend the data. Table 10.2 provides a comparison of the BLE PHY variants.

We note that for all variants the maximum application data rate is lower than the protocol data rate (Sect. 10.4.4). This is due to the limit on the number of packets per connection interval, delay between packets, packet overhead, etc.

10.3.4 Bit Stream Processing

Bit stream processing for BLE PHY variants LE M1 and LE M2 is as shown in Fig. 10.3, and for LE Coded $S = 2$ and LE Coded $S = 8$ as shown in Fig. 10.4. The processes employed are:

- **Encryption/decryption**: Encryption, which is not applied in all situations, is a process of scrambling incoming data so that only parties with a decryption key can unscramble the data at the receive end.
- **CRC generation/CRC checking**: CRC (Sect. 5.2) is employed for error detection. It is calculated on the PDU fields of all LL packets (Sect. 10.4.2) and is always calculated after encryption.
- **Whitening/dewhitening**: This is scrambling (Sect. 8.6.4.3) and is used to avoid long sequences of zeros and ones.
- **FEC encoding/decoding**: This is BCC encoding /decoding as discussed in Sect. 10.3.3 above.
- **Pattern Mapper/de-mapper**: This is used to vary the coding effective coding strength and rate and is as discussed in Sect. 10.3.3.

10.3.5 Time Division

BLE employs half-duplex communication, that is, it has the ability to transmit and/ or receive but not both at the same time. However, all BLE PHY variants operate in a time division duplex (TDD) mode where from one very short instant to the next transmission direction can be reversed leading to the appearance of full-duplex operation.

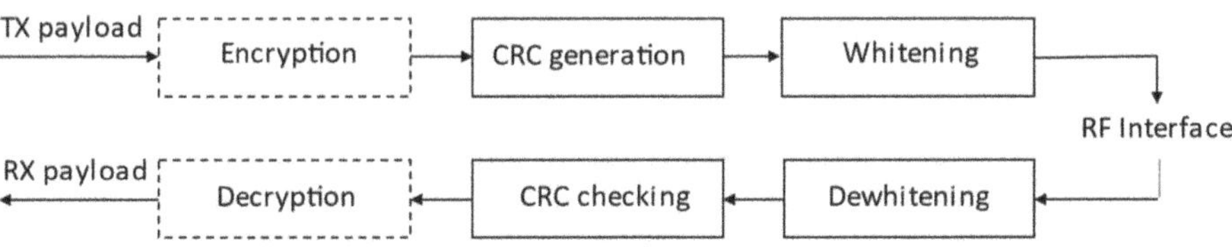

Fig. 10.3 Bit stream processing for LE M1 and LE M2

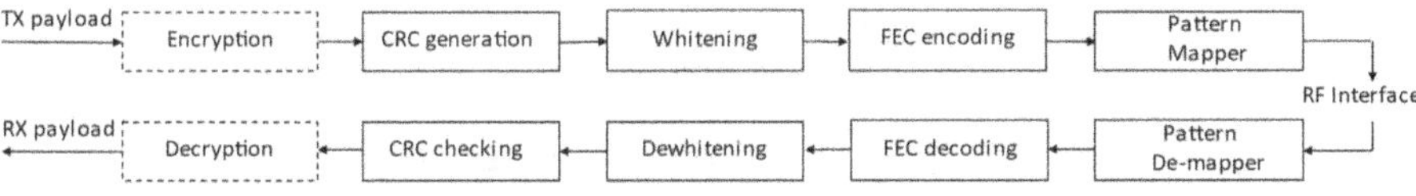

Fig. 10.4 Bit stream processing for LE coded $S = 2$ and LE coded $S = 4$

10.3.6   Transmitter Power and Receiver Sensitivity

The maximum *transmitter output power* for BLE is specified to lie between −20 dBm
(0.01 mW) and +20 dBm (100 mW). This is the power level at the antenna connector.

   BLE *receiver sensitivity* is defined as that receiver input level at which a specified bit error rate (BER) is achieved and is a function of the maximum supported
packet payload length in bytes (Sect. 10.4.2). It varies from 0.1% for lengths of
1–37 bytes to 0.017% for lengths of 128–255 bytes. Furthermore, the specification
mandates that the actual sensitivity must be a minimum of −70 to −82 dBm depending on the PYH used. Specifically, for LE 1M and LE 2M, it is −70 dBm, for LE
Coded $S = 2$ it is −75 dBm, and for LE Coded $S = 8$ it is −82 dBm. In reality, BLE
receivers typically achieve sensitivity of −95 dBm or better.

10.3.7   Antenna Switching

BLE, via capability added in Core Specification V5.1, provides two methods to calculate the direction from which a received signal is transmitted, namely *Angle of Arrival*
(AoA) method and *Angle of Departure* (AoD) method. Both techniques require that
one of the two communicating devices have an array of multiple antennas. With the
AoA method, the array is in the receiving device and involves a process of the receiver
switching from one antenna to another during reception. With the AoD method, the
array is in the transmitting device and involves a process of the transmitter switching
from one antenna to another during transmission. Direction finding signals are part of
standard BLE packets in the form of the Constant Tone Extension field (Sect. 10.4.2).
Switching takes place during time periods labeled switch slots and is carried out
according to a prescribed pattern. Figure 10.5 illustrates AoA and AoD methods.

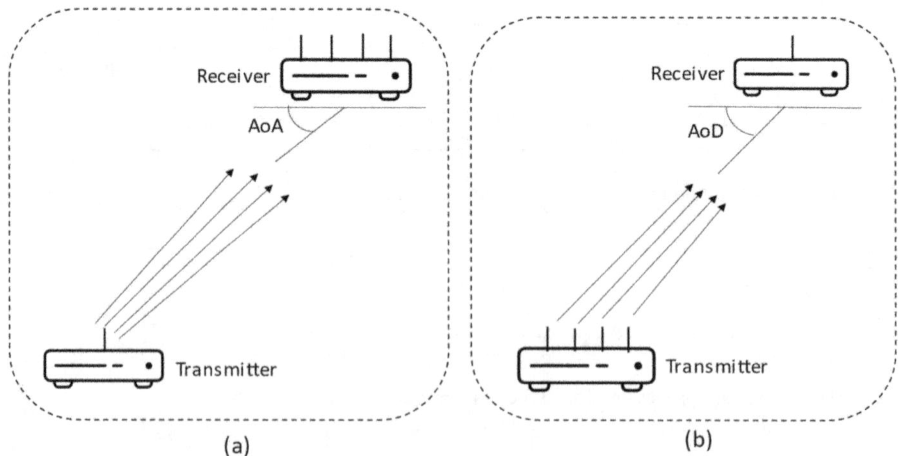

Fig. 10.5 (**a**) Angle of Arrival (AoA) and (**b**) Angle of Departure (AoD) methods

10.4   The Link Layer (LL)

10.4.1   Overview

The LL has several functions, these being somewhat similar to the MAC layer in 5G NR and the Data Link layer in 802.11. It can create many types of packets that are transmitted over the air via the physical layer with which it interfaces directly. It is responsible for scheduling, RF channel selection, packet formats, error checking, etc. It supports connected communication, i.e., the establishment of a two-way connection link before any data can be transferred, as well as connectionless communication, i.e., no need to establish a two-way connection link before data can be transferred.

In BLE, the physical channel is divided into time units called events. Data is communicated between BLE devices in packets that are located in these events. With BLE, there are two types of events, namely *advertising events* and *connection events*. With Advertising, a BLE advertiser (defined below) broadcasts packets to every device around it. With a connection, both the master in a Central Role (defined below) and a slave in a Peripheral Role (defined below) send packets to each other.

The LL specification defines a number of device operating states but only permits one state to be active at a time. Such states and their relationship to each other is shown in Fig. 10.6.

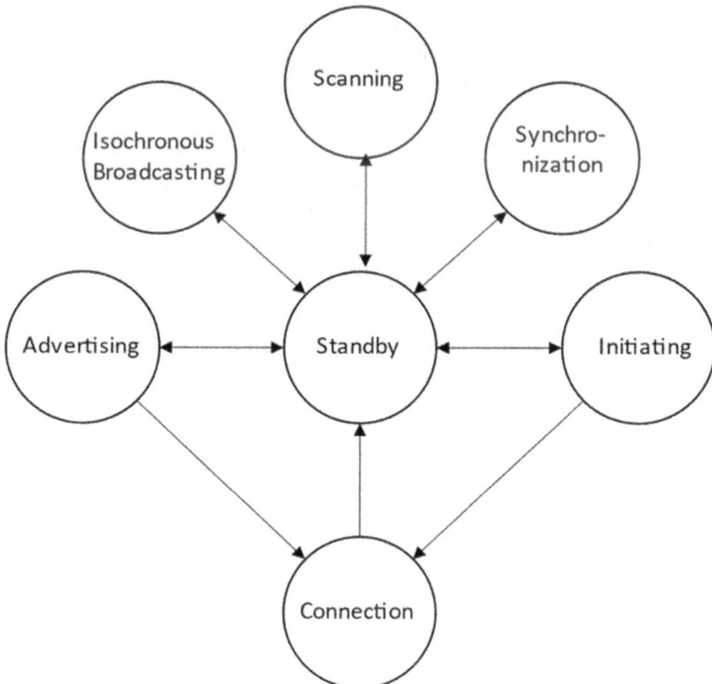

Fig. 10.6 Link Layer defined states

In the **Standby** state, a device does not transmit or receive any packets. It can be entered from any other state.

In the **Advertising** state, a device transmits advertising packets in advertising events and possibly listens to and responds to responses generated by its outgoing packets. This state can be entered from the standby state and a device in it is called an *advertiser*.

In the **Scanning** state, which can be entered from the Standby state, a device listens for advertising packets from advertisers. A device in this state is known as a *scanner*.

In the **Initiating** state, a device wants to form a connection to another specific device and listens for connectable advertising packets. When it detects advertising packets from the specific device, it transmits a connection request to the advertiser which in turn sets up a point-to-point connection between the two devices. This connection allows both devices to communicate with each other over the physical channel. A device in this state is known as an *initiator*, and this state can be entered from the Standby state.

In the **Connection** state, which can be entered from the Initiating or Advertising state, a device is said to be in a connection. Two roles are defined within this state, namely *Central Role* and *Peripheral Role*.

When entered from the Initiating state, a device in the Connection state is said to be in the Central Role. A device in this role communicates with a device in the Peripheral Role and defines the timing of transmissions. It is referred to as a *master* device and is typically a smartphone, or a tablet, or a PC.

When entered from the Advertising state, a device in the Connection state is said to be in the Peripheral Role. A device in this role is referred to as a *slave* device and communicates with a single central device.

(A network composed of a master and one or more slave with point-to-point connections between the master and each slave is referred to as a *piconet*.)

In the **Synchronization** state, which can be entered from the Standby state, a device listens for physical packets coming from a specified device that form a specific periodic advertising train. A device in this state is referred to as a *Synchronized Receiver*.

In the **Isochronous** Broadcasting state, which can be entered from the Standby state, a device transmits isochronous data packets on an isochronous physical channel. A device in this state is known as an *Isochronous Broadcaster* (Sect. 10.4.7). Isochronous data is data that is transmitted a constant rate, not in a start stop fashion as is the case with asynchronous data. Isochronous transmission is advantageous to transmission of time-bound data such as voice and video.

10.4.2 *Packets*

The LL defines two packet types, one used by the two uncoded PHYs, LE 1M and LE 2M, and the other by the two coded PHYs, LE Coded $S = 1$ and LE Coded $S = 8$.

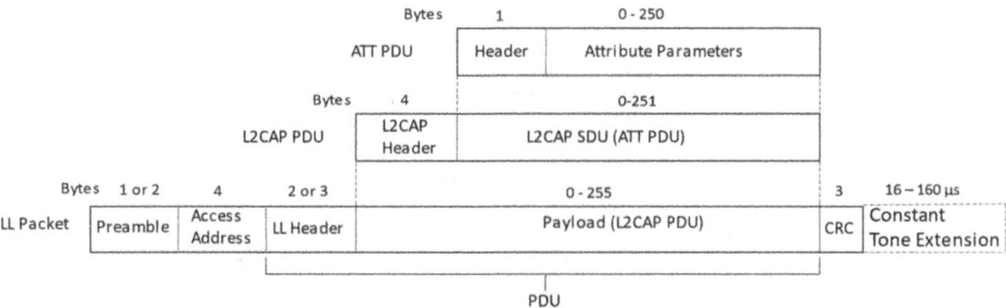

Fig. 10.7 ATT, L2CAP, and Link Layer packet format for LE uncoded PHYs

To get a sense of downward data flow through the protocol stack, a simplified example of such flow from the ATT protocol level to the LL level for uncoded PHYs is shown in Fig. 10.7.

At the top, we see the ATT PDU. It consists of a 1-byte Header called Attribute Opcode. This is followed by Attribute Parameters, followed in turn by 0–12 byte optional Authentication Signature (not shown). The Attribute Parameters size is variable, but the specified maximum length of an Attribute is 255 bytes.

Following the ATT PDU, we have the L2CAP PDU. The theoretical maximum L2CAP SDU is 65, 535 bytes. However, it shall not be greater than the peer device's Maximum Transmission Unit (MTU) for the channel and implementations are required to support a connectionless MTU of 48 bytes.

At the bottom, we have the LL Packet for LE uncoded PHYs. Following is a description of the various fields:

- **Preamble**: The preamble is used in the receiver to perform frequency synchronization, symbol timing estimation, and automatic gain control training. It's a fixed sequence of alternating 0 and 1 bits. For LE 1M it is 8 bits (1 byte) and for LE M2 it is 16 bits (2 bytes).
- **Access Address**: The access address, which has a 32 bit (4 byte) value, is used in the receiver to differentiate received signals from background noise and to determine the relevance of a packet, ignoring packets where the access address is not relevant to it.
- **PDU:** The PDU is either an Advertising Physical Channel PDU or a Data Physical Channel PDU. It consists of the **LL Header** followed by the **Payload** which, in the case of Data Physical Channel PDUs, in turn may be followed by a 4-byte **Message Integrity Check (MIC)** field (not shown).
- **LL:** The LL Header is 2 bytes long unless, in the case of Data Physical Channel PDUs, the Constant Tone Extension is present, in which case it is 3 bytes long.
- **Payload:** The payload is the L2CAP PDU from the L2CAP layer above.
- **MIC:** The MIC field is 4 bytes long and is applied when the PDU field is encrypted so as to protect the PDU from being tampered with.
- **CRC:** The CRC is 3 bytes long and is calculated over the PDU.
- **Constant Tone Extension:** The Constant Tone Extension is optional and of variable length. It consists of a constantly modulated series of 1s. Its purpose is to enable Angle of Arrival (AoA) and Angle of Departure (AoD) direction finding.

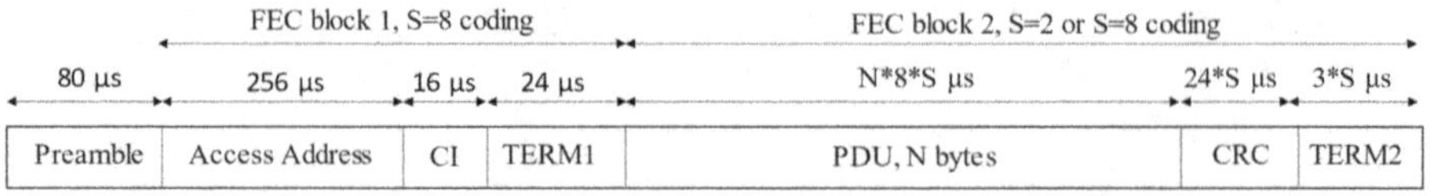

Fig. 10.8 Link Layer packet format for LE Coded PHYs

The optionally encrypted PDU field and the CRC field are subject to whitening (see Fig. 10.3) before the packet is transmitted.

Figure 10.8 shows the LL packet format for LE Coded PHYs.

Following is a description of the various fields in the coded PHYs:

- **Preamble**: Used for the same purposes as in the uncoded packets. It is 80 symbols in length and hence 80 µs long. It consists of 10 repetitions of the symbol pattern 00111100. The Preamble is not coded.
- **FEC block 1**: This block consists of the **Access Address**, the **Coding Indicator (CI)**, and **TERM 1**. All these fields use the $S = 8$ coding scheme.
- **Access Address**: The access address is as described for the uncoded PHYs. Note, however, that as the $S = 8$ coding scheme is applied, the 32-bit value in the uncoded state is now expanded to $32 \times 8 = 256$ bits and hence a field length of 256 µs.
- **Coding Indicator**: The coding indicator consists of 2 bits prior to coding and indicates if the coding on FEC block 2 is $S = 2$ or $S = 8$.
- **TERM 1 and TERM 2**: Each terminator sequence at the end of each FEC block is 3 uncoded bits long and its purpose is as described in Sect. 10.3.3.
- **FEC block 2**: This block consists of the **PDU**, **CRC**, and **TERM 2** fields and uses either the $S = 2$ or $S = 8$ coding scheme depending on the value of the CI.
- **PDU**: The PDU is as described above for the uncoded PHYs. It consists of N uncoded bytes leading to $N * 8 * S$ coded bits and hence a length of $N * 8 * S$ µs.
- **CRC**: The CRC is 24 uncoded bits long and hence $24 * S$ µs long.

As with the uncoded PHYs, the optionally encrypted PDU field and the CRC field are subject to whitening prior to encoding (see Fig. 10.4) before the packet is transmitted.

10.4.3   *Connection State and Associated Timing Parameters*

A device can establish a connection with an advertising device by responding to a packet received from the advertising device with a PDU that requests a connection. The device requesting the connection moves from the Standby state to the Initiating state and then to the Connection state where it assumes the Central role. In this role, it is referred to as a Central or a Master. The other device moves from the Advertising state to the Connection state and assumes the Peripheral role and in this role is referred to as a Peripheral or a Slave.

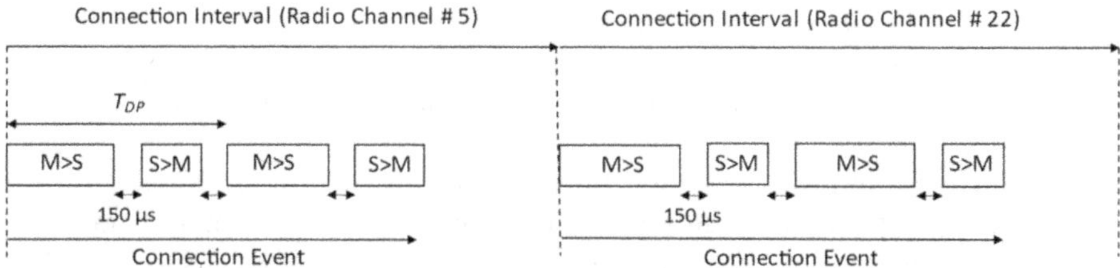

Fig. 10.9 Basic packet exchange during connection intervals

When a connection is established, it is defined for a specified interval of time referred to as the *Connection Interval*. This interval can range from 7.5 ms to 4 s in multiples of 1.25 ms and exists over one radio channel, i.e., there is no hopping during the interval. During the interval, packets flow from the Master to Slave and vice versa. The time between two consecutive packets in the same connection Interval is referred to as the *Inter Frame Space* (IFS) and is specified as 150 µs. The time that packet flow exists within a Connection Interval is referred to as the *Connection Event* and need not occupy the full Connection Interval. Figure 10.9 shows basic packet exchange during two Connection Intervals. The format of the packets shown in this figure is either as depicted in Fig. 10.7 or 10.8. The Master to Slave (M > S) packets shown are assumed to be data carry ones whereas the Slave to Master (S > M) packets shown are assumed to be acknowledgment packets and thus have 0 bytes in the payload field, the designation as being acknowledgment packets being coded in the PDU header.

We note that for Apple iPhones, the minimum Connection Interval is 15 ms, whereas for Android based phones it is 7.5 ms.

10.4.4 Approximate Maximum Application Data Rates

With BLE all symbols are modulated with GFSK which is a binary modulation scheme. Thus, the symbol rate always equals the modulating signal bit rate. This bit rate, however, is not the rate that application data is transmitted at. The actual application data rate depends on a number of factors including:

- The symbol rate.
- The coding applied.
- The percentage of the Connection Interval occupied by LL packets.
- The interframe space (IFS).
- The header overhead added to the application data.

For maximum data rate the Connection Interval must be fully utilized. Shown in Fig. 10.9 is a time interval labeled T_{DP}. This time represents the time consumed to transfer one data packet from a master to a client or vice versa, receive and acknowledgment, and be ready for transmission of another data packet. To fully utilize the

Connection Interval, the Connection Interval must be an integer multiple of T_{DP}. When this is the case, and assuming the M > S and S > M periods don't change, then the data rate achieved in the period T_{DP} will be the data rate achieved over the whole Connection Interval.

Our interest is in the maximum application data rate. Within the period T_{DP}, the maximum application data transferred will be most of the maximum data in the Attribute Parameters, this maximum being 250 bytes or 2000 bits (See Fig. 10.7). Since these bits are transferred every T_{DP} seconds, then the Maximum application data rate, R_m, is $2000/T_{DP}$.

Simple arithmetic calculations will show that:

- **For LE 1M** LL packets, the length of a LL packet with a maximum payload of 255 bytes is 2120 µs, the length of an acknowledgment packet with a zero-byte payload is 80 µs, and thus $T_{DP} = 2120 + 150 + 80 + 150 = 2500$ µs. As a result:
- $R_m = 2000/2500$ µs $= 800$ kb/s.
- **For LE 2M** LL packets, the length of a LL packet with a maximum payload of 255 bytes is 1064 µs, the length of an acknowledgment packet with a zero-byte payload is 44 µs, and thus $T_{DP} = 1064 + 150 + 44 + 150 = 1408$ µs. As a result:
- $R_m = 2000/1408$ **µs = 1.42 Mb/s.**
- **For LE Coded $S = 2$** LL packets, the length of a LL packet with a maximum payload of 255 bytes is 4510 µs, the length of an acknowledgment packet with a zero-byte payload is 430 µs, and thus $T_{DP} = 4510 + 150 + 430 + 150 = 5240$ µs. As a result:
- $R_m = 2000/5240$ **µs = 380 kb/s.**
- For LE Coded $S = 8$ LL packets, the length of a LL packet with a maximum payload of 255 bytes is 16,912 µs, the length of an acknowledgment packet with a zero-byte payload is 592 µs, and thus $T_{DP} = 16,912 + 150 + 592 + 150 = 17,80$ 4 µs. As a result:

- $R_m = 2000/17,804$ **µs = 110 kb/s.**

It should be noted that although the symbol rate is doubled from LE 1M to LE 2M the maximum application data rate only increases by 78%, this being largely due to the invariant size of the interframe space.

10.4.5   Adaptive Frequency Hopping

Bluetooth Low Energy (BLE) accesses the 37 General Purpose channels of the 2.4-GHz band by the application of frequency hopping spread spectrum (Sect. 6.5). It does this using a technique called *adaptive frequency hopping* that minimizes the negative effect of interference by seeking as it hops clear transmission channels and thus avoiding packet collisions. Channels that are busy or noisy, referred to as *unused* channels, are tracked dynamically, and as it hops it avoids these channels, using instead the *used* channels, i.e., those channels that are not busy or noisy.

As seen in Sect. 10.4.3 above, at each connection event, a pair of connected devices are able to exchange packets at precisely timed Connection Intervals. Frequency change occurs at the start of each such interval with the newly assigned radio channel being deterministically chosen from the set of used channels. Each connected device switches to the selected channel and as transmission moves from one connection event to another communication takes place using only used channels thus significantly minimizing the probability of collisions occurring. Used channels to hop to are chosen using a channel selection algorithm and a table of data called the *channel map* that is cognizant off all used and unused channels.

Over time, the channel map may change. All channels are constantly monitored and as the condition on each channel changes the channel amp is updated. In this way, the channel selection algorithm constantly adapts to the channel conditions currently being experienced and thus maximizes the probability of collision free performance. The master device in a connection performs channel monitoring, maintains the channel map, and shares this with the slave device. The slave can also perform channel monitoring, send its findings to the master, and the master can take into account these findings in maintaining the channel map.

The Connection interval determines the hop rate. As the Connection Interval varies between 7.5 ms and 4 s, the hop rate varies between $1/(0.0075) = 133.33$ hops per second seconds and $1/4 = 0.25$ hops per second.

Figure 10.10 shows the 40 Bluetooth 2-MHz channels, where Wi-Fi channels 1 and 11 are present thus creating unused channels. A frequency hopping sequence from Channel # 11 to Channel # 21, then on to Channel # 36 is shown, this sequence avoiding unused channels.

10.4.6   Extended Advertising

Prior to Bluetooth Core Specification Ver. 5.0, advertising, now referred to as *Legacy advertising*, used only the three primary advertising channels, namely channel RF Channel numbers 0, 12, and 39 (Index numbers 37, 38, and 39). However, Core Specification Ver. 5.0 introduced *Extended advertising* which uses all 40 Bluetooth LE channels for advertising. With Extended advertising, much larger amounts of

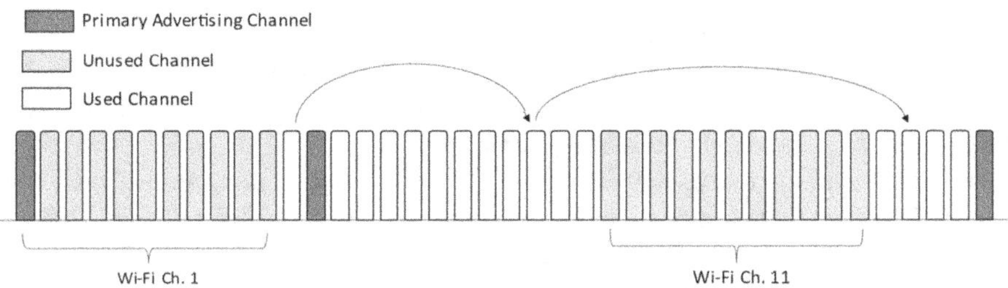

Fig. 10.10 Adaptive frequency hopping

data can be broadcast and advertising can be performed to a deterministic schedule. It results in better spectral efficiency and significant improvement in duty cycle and channel contention.

With extended advertising, the primary advertising channels carry only header data, and the 37 secondary advertising channels therefore carry most of the data. With Legacy advertising, the same data is transmitted up to three times on the three different primary advertising channels. With Extended advertising, however, payload data is transmitted only once, with small headers on the primary channels referencing it. Thus, the total amount of data transmitted is less than the similar case using legacy advertising resulting in a reduction in the effective duty cycle. Furthermore, with advertising data able to use all available channels, and only small header packets using the primary channels, there is less contention on the primary channels.

Extended advertising supports a concept called *periodic advertising* where advertising packet transmission takes place at precisely timed intervals and where a method is provided for observing devices to determine the advertising devices periodic advertising schedule and then synchronize their scanning with it.

10.4.7 *Isochronous Communications and LE Audio*

Isochronous data is data that is transmitted a constant rate, not in a start stop fashion as is the case with asynchronous data. Isochronous transmission is advantageous to transmission of time-bound data such as voice and video. LE isochronous communication was introduced in Core Specification Version 5.2. Here, in one application, data may be transmitted from a central device to peripheral devices in isochronous streams which belong to isochronous groups. Peripherals wait until all streams that are members of the same group have delivered their packets before they all process these packets in a synchronized fashion.

Two forms of isochronous LE physical channel data streams are defined. One is the *Connected Isochronous Stream* (CIS) which uses connection-oriented communication and supports the bidirectional transfer of data. The other is the *Broadcast Isochronous Stream* (BIS) which uses connectionless communication and provides unidirectional data communication.

A single CIS stream provides point-to-point isochronous communication between two connected devices and may use any of the Bluetooth LE physicals. Bidirectional communication is supported, and an acknowledgment protocol is used. CIS streams exist within groups called *Connected Isochronous Groups* (CIGs) which may contain a maximum of 31 CISs.

CISs have timing events that occur at a regular interval ranging from 5 ms to 4 s. Each CIG event that encompasses sequentially all the associated CISs starts at the anchor point of the earliest CIS in the group. Each CIS event is broken down into one or more subevents and the RF channel is changed at the start of each subevent. In a CIS subevent, the Central transmits once and the Peripheral responds.

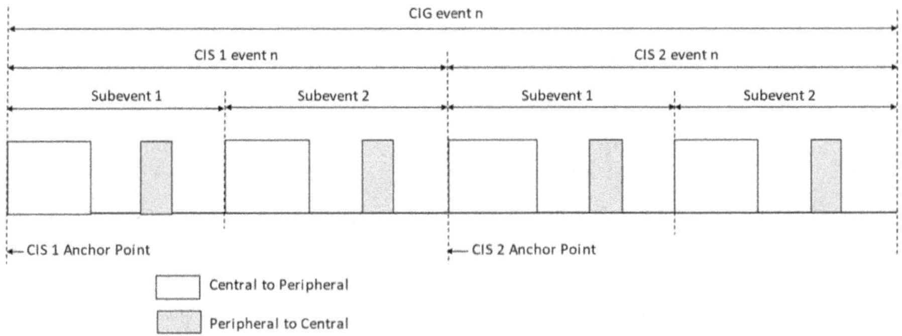

Fig. 10.11 A CIG event which includes two CIS events, with each CIS event including two subevents

Figure 10.11 shows a CIG event which includes two CISs with each CIS event including two subevents.

A Broadcast Isochronous Stream (BIS) provides broadcast isochronous communication between one transmitter and many receiving devices and, like CIS, may use any of the Bluetooth LE physicals. BIS streams exist within groups called *Broadcast Isochronous Groups* (BIGs) which may contain a maximum of 31 BISs. Only unidirectional communication is supported by a BIS and an acknowledgment protocol is not used. As discussed in Sect. 10.4.1, one of the LL states is the Isochronous Broadcasting state that was added in the Core Specification Version 5.2. It is in this state that LE BIS communication takes place, and the broadcasting device is referred to as an *Isochronous Broadcaster*.

For each BIS within a BIG, there exists a schedule of transmission time slots known as events. Each event is divided into one or more subevents. During a subevent, a single packet is transmitted by the broadcaster and the RF channel is changed at each subevent. Communication is unidirectional, there being no acknowledgment. Figure 10.12 shows two BIG events each of which includes two BISs with each BIS including two subevents. Such a scheduling structure would exist if two streams, one stereo left and one stereo right, were being broadcast to multiple Peripherals.

LE isochronous communication was primarily designed to support LE Audio, the latest generation of Bluetooth audio. LE Audio is preceded by Classic Audio which operates on the Classic physical layer. LE Audio brings a number of improvements relative to Classic Audio including a new more effective audio codec, multi-stream audio, hearing aid support, and broadcast audio sharing.

The new high quality, low power audio codec used with LE Audio is called LC3 which is short for "Low Complexity Communications Codec." Compared to the "Subband Coded" (SBC) used in Classic Audio, it is much more efficient in the processing and delivery of an audio output while supporting a wide range of sample rates, bit rates, and frame rates.

With multi-stream audio, it is possible to transmit multiple, independent, audio streams from an audio source, such as a smartphone, to one or more audio receiving

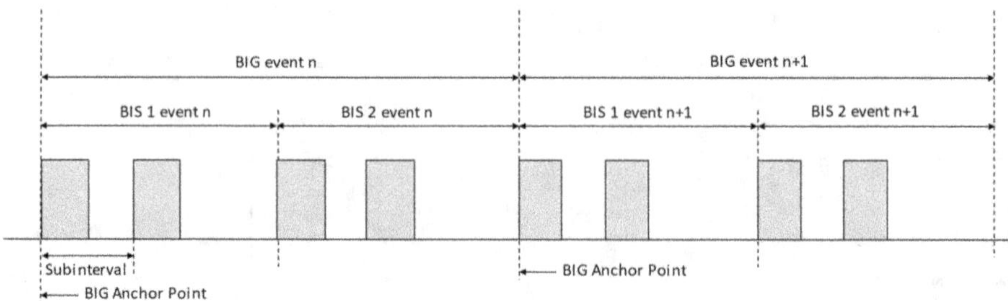

Fig. 10.12 BIG events which includes two BISs each, with each BIS including two subevents

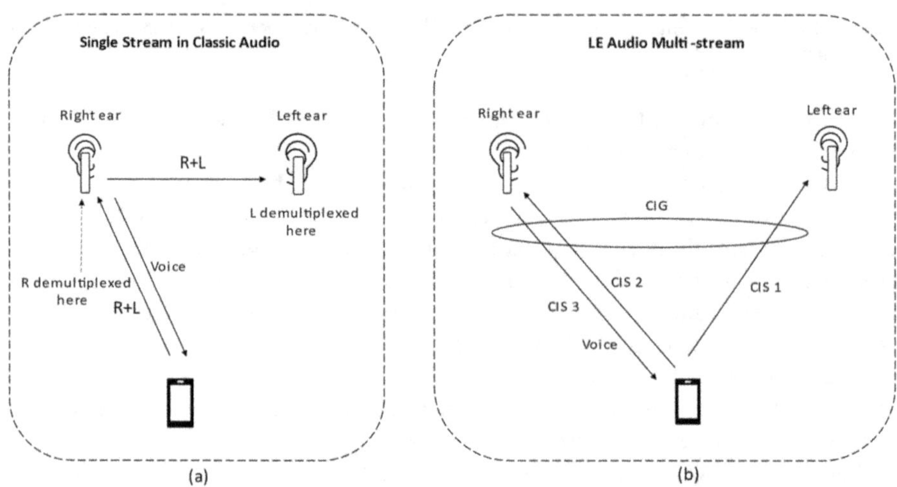

Fig. 10.13 (**a**) Classic audio single stream, (**b**) LE Audio multi-stream

devices such as earbuds or earphones and synchronize the received data utilizing LE CISs within a CIG. Classic Audio supports only single point-to-point audio streams. Thus, to serve a stereo headset or ear bud, for example, it must transmit a multiplexed right and left channel signal to one receiver and resend this multiplexed signal on to the other receiver, demultiplexing the right signal in the right receiver and the left signal in the left receiver. Figure 10.13 shows ear buds operating with (a) Classic Audio and (b) LE Audio.

Support of Bluetooth technology in hearing aids based on Classic Audio has been based on proprietary implementations. LE Audio, however, allows the development of standard Bluetooth hearing aids which should increase to availability of Bluetooth enabled hearing aids. LE Audio enabled hearing aids will likely be much more power efficient than Classic Audio enabled ones given the efficient power consumption of Bluetooth LE and the LC3 codec.

LE Audio supports broadcast audio, Auracast™, enabling an Isochronous Broadcaster to broadcast one or more audio streams to an unlimited number of audio receiving devices utilizing BISs within a BIG.

10.5   Advertising Coding Selection

Among four new features made available with Bluetooth 5.4 is one called *Advertising Coding Selection*. We recall that two of BLEs physical variants, namely LE coded $S = 2$ and LE coded $S = 8$ are convolution coded, where the i parameter, of value 2 or 8, determines how much coding is applied. Prior the Bluetooth 5.4, it was not possible to specify the value of S which was to be used with extended advertising. With Advertising Coding Selection, the host is able to select the value of S that the controller uses for transmitting extended advertising PDUs. This capability can improve a situation where, for example, a device receives from a peer device $S = 8$ coded PDUs but responds with $S = 2$ coded PDUs which, as we know, have a lower range than $S = 8$ coded ones. As a result, the peer device is out of range to receive the $S = 2$ coded PDUs. Here, the host can instruct the controller to transmit with $S = 8$ PDUs and thus bring the peer into range.

10.6   Bluetooth Version 6.0

In September 2024, Bluetooth Special Interest Group (SIG) released Bluetooth Version 6.0 [2]. This release includes the following new features and enhancements that collectively represent a significant improvement in technology:

- **Bluetooth Channel Sounding**: let us one determine with high precision the range and direction between two Bluetooth devices.
- **Decision-Based Advertising Filtering**: improves scanning efficiency by allowing a scanning device to use the content of a packet received on a primary advertising channel to conclude if it should scan for related packets on secondary channels. This approach reduces the time spent scanning on secondary channels for packets that may not contain relevant PDUs, thus improving scanning efficiency.
- **Monitoring Advertisers**: here the host component of an observer device may direct the Bluetooth LE controller (see Sect. 10.2) to filter duplicate advertising packets. When such filtering is active, the host only receives a single advertising packet from each unique device, improving the host efficiency.
- **ISOAL Enhancement**: here the Isochronous Adaptation Layer (ISOAL) (see Sect. 10.2) has been improved by specifying a new PDU framing mode that reduces latency and improves reliability for situations that are particularly sensitive to latency issues.
- **LL Extended Feature Set**: this feature allows devices to exchange information about a larger number of LL features that they each support than previously supported.
- **Frame Space Update**: Pre-Version 6.0 specifications specified a constant value of 150 µs for the time separating adjacent transmissions of packets in a connection event or connected isochronous stream. In Version 6.0, this time is now

negotiable and may be longer or shorter than 150 µs. Shorter frame spacing can improve throughput for LE Audio devices, fitness trackers, etc.

10.6.1 Bluetooth Channel Sounding

Bluetooth Channel sounding is arguably the most significant feature of Bluetooth 6.0, likely to revolutionize distance and directional measurement accuracy between two Bluetooth devices. In this section an overview is provided at a high level of the technology behind Bluetooth Channel Sounding, based largely on material in Sect. 3 of [4].

In pre-Version 6.0 releases, distance calculation between devices could only be achieved via a method referred to as *path loss calculation*. Here, the receiving device measures the received signal strength, P_r say, and knowing the transmitter output power, P_t say, the path loss $P_t - P_r$ can be calculated. Then, using Eq. (3.2), the path distance is easily calculated. The limitation of this method is that the accuracy of the distance calculated is inversely proportional to the distance. For example, for a transmission frequency of 2401 MHz and a distance of 100 m, the true path loss is 80.01 dB. Should the computed loss be 80.21 dB (off by 0.2 dB), then the calculated distance would be 102.3 m, an error of 2.3 m. For the same frequency and a distance of 500 m, the true path loss is 93.99 dB. Should the computed loss be 94.19 dB (again off by 0.2 dB), the calculated distance would be 508 m, an error now of 8 m. In addition to this distance related accuracy variability, this method is also vulnerable to interference and other environmental factors. Because of these weaknesses, SIG introduced in Version 6.0 Bluetooth Channel Sounding, a more accurate approach to distance finding. Bluetooth Channel Sounding comprises two distinct methods of distance measurement that may be used independently or together. These two methods are called *phase-based ranging* (PBR) and *round-trip timing* (RTT).

10.6.1.1 System Architecture and RF Channels

Channel Sounding occurs between two connected Bluetooth devices. The device that desires to calculate the distance between itself and another device is called the *Initiator*. The other device is called the *Reflector*. During channel sounding several two-way exchanges take place between the Initiator and Reflector during predetermined time slots.

Regarding the role in the protocol stack, which is shown in Fig. 10.1, we note that channel sounding is primarily a function of the Controller as opposed to the Host. The Initiator's Controller feeds low level measurements up the stack to the Application layer via the Host where this data is used to calculate the distance.

Devices that employ channel sounding may incorporate an antenna array. With an array, the accuracy of distance measurements can be improved by reducing the impact of multipath as a result of improved antenna directivity.

For channel sounding purposes, 72 channels in the 2.4-GHz band are specified, each with 1 MHz rather than the usual 2-MHz bandwidth and a unique channel index value. These channels are arranged to ensure that the LE primary advertising channels are avoided.

10.6.1.2   Phase-Based Ranging

The phase-based ranging (PBR) distance measurement method utilizes the phase of a radio signal and its relationship with that signal's frequency and wavelength. When employing this method, the Initiator transmits a series of CS Tones of frequency f_1 say to the Reflector. A *CS Tone* is a signal that uses amplitude shift keying (ASK) to create a symbol whose frequency is fixed. On receiving this signal, the Reflector retransmits it back to the Initiator, which, on receiving it, measures its phase, P_{f_1} say. See Fig. 10.14. Next the Initiator transmits another unmodulated signal, but this time of a different frequency, f_2 say. As before, the Reflector retransmits it back to the Initiator which on receiving it measures its phase, P_{f_2} say, which is different to P_{f_1} because of the change in frequency. The Initiator is now able to calculate the distance d between the two devices by applying a formula that includes the frequency difference $(f_1 - f_2)$, the phase difference $\left(P_{f_1} - P_{f_2}\right)$, and the speed of light.

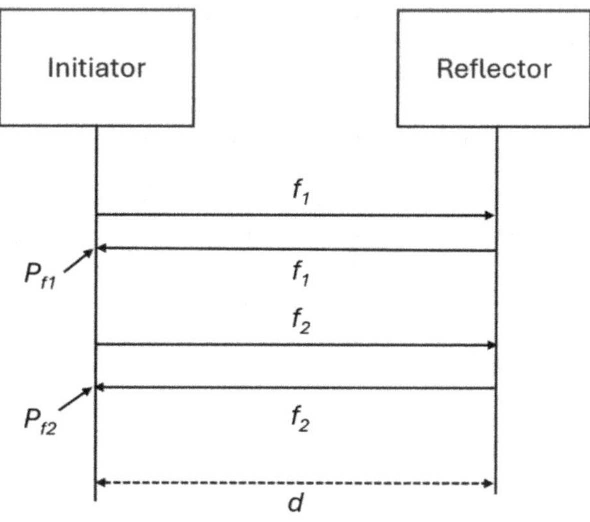

Fig. 10.14 Phase-based ranging with RF signals

As indicated above, multipath propagation can negatively impact ranging accuracy. However, collecting phase measurements from an increasing number of frequencies can minimize this problem. Thus, in reality, the devices will likely exchange more than two signals, each on a different frequency, as the more measurements are made to more accurate will be the result.

The above PBR explanation is somewhat simplified but does convey the essence of the process.

10.6.1.3   Round-Trip Timing

With the Round-Trip Timing (RTT) method of distance estimation, the Initiator transmits a series of *CS Sync packets* to the Reflector. CS Sync packets can be transmitted using LE 1M, LE 2M, or LE 2M 2BT (see Sect. 10.6.1.4) physical variants. On receipt of these packets, the reflector sends them back to the Initiator.

To facilitate the calculation of the distance d between the Initiator and Reflector, the Initiator records the time that the transmitted packet departs, this time being known as Time of Departure or ToD. When the return packets are received from the Reflector, the Initiator records this time, this time being known as the Time of Arrival or ToA.

If no time were to be lost as the Reflector turns around the signal, then the distance between the devices d could be easily calculated by multiplying the total transmission time ToA-ToD, T_{A-D} say, by the speed of light and dividing by two (as T_{A-D} is the time for a round trip). In reality, however, time is lost as the Reflector turns around the signal. In Bluetooth RTT channel sounding, this time at the reflector, T_R say is known. Thus, the true distance is calculated by multiplying $(T_{A-D} - T_R)$ by the speed of light and dividing by 2.

10.6.1.4   The LE 2M 2BT PHY

Prior to Version 6.0, Bluetooth physical variants were, as given in Table 10.2, LE 1M, LE 2M, LE Coded $S = 2$, and LE Coded $S = 8$. In Version 6.0, a new physical layer configuration called LE 2M 2BT has been introduced which may be used only with Bluetooth Channel Sounding.

Key features of LE 2M 2BT are:

- Modulation: GFSK.
- Symbol rate: 2 Mb/s.
- Bandwidth-bit period product BT: 2.0.

- Minimum frequency deviation: 420 kHz.
- Error detection: Not applicable.
- Error correction: Not applicable.
- Requirement: Optional. To be used only with Channel Sounding.

For all the pre-Version 6.0 variants, BT was 0.5. The higher BT of LE 2M 2BT results in a squarer, narrower, and hence shorter symbol span. Certain Bluetooth Channel Sounding security attacks could involve manipulating radio symbols in order that the target device incorrectly calculates the distance to the other associated device. Because the LE 2M 2BT has a shorter symbol span, symbols are harder to intercept and manipulate in this fashion.

10.7   Comparison Between Bluetooth LE and Bluetooth Classic

Above key features of Bluetooth LE (Versions 5 and 6) have been presented. With an understanding of these features, a comparison with key features of Bluetooth Classic, the predecessor technology, is in order. Such a comparison is presented in Table 10.3.

Table 10.3 Comparison of key features between Bluetooth Classic and Bluetooth LE

Feature	Bluetooth LE (ver. 5 and 6)	Bluetooth Classic (BR/EDR)
Frequency band	2.4-GHz ISM band	2.4-GHz ISM band
Channels	40 channels with 2-MHz spacing	79 channels with 1-MHz spacing
Channel usage	Adaptive frequency hopping spread spectrum	Frequency hopping spread spectrum
Modulation	GFSK	GFSK, $\pi/4$ DQPSK, 8DPSK
Max data rate: Theoretical/application	LE 2M: 2 Mb/s/1.42 Mb/s LE 1M: 1 Mb/s/0.8 Mb/s LE coded ($S = 2$): 0.5 Mb/s/0.38 Mb/s LE coded ($S = 8$): 0.125 Mb/s/0.11 Mb/s	EDR: 3 Mb/s/2.1 Mb/s EDR: 2 Mb/s/1.4 Mb/s BR: 1 Mb/s/0.7 Mb/s
Max. transmission power	100 mW (20 dBm)	100 mW (20 dBm)
Rx sensitivity	LE 2M: ≤ -70 dBm LE 1M: ≤ -70 dBm LE coded ($S = 2$): ≤ -75 LE coded ($S = 8$): ≤ -82	≤ -70 dBm
Network topologies	Point-to-point (including piconet), broadcast, mesh	Point-to-point (including piconet)
Location and direction finding	Supported	Not supported
Audio stream	Multi-stream	Single stream

10.8 Summary

In this chapter, emphasis was placed on the latest two versions of Bluetooth, namely versions 5 and 6, both of which fall under the nomenclature BLE. A key feature of BLE is that it is highly efficient in its use of power. We examined, for BLE, protocol architecture, the RF spectrum supported, the physical layer and LL, and the key technologies employed therein, and finally a comparison between BLE and Bluetooth Classic was presented.

References

1. SIG Core Specification Working Group (2023) Bluetooth core specification, version 5.4, Bluetooth Special Interest Group, Kirkland, Washington
2. SIG Core Specification Working Group (2024) Bluetooth core specification, version 6.0, Bluetooth Special Interest Group, Kirkland, Washington
3. SIG Core Specification Working Group (2023) Bluetooth core specification, version 5.4, technical overview, Bluetooth Special Interest Group, Kirkland, Washington
4. SIG Core Specification Working Group (2024) Bluetooth core specification, version 6.0, feature overview, Bluetooth Special Interest Group, Kirkland, Washington
5. Martin Woolley (2024) The Bluetooth low energy primer, version 1.2.0, Bluetooth Special Interest Group, Kirkland, Washington

Index

GPSR Compliance
The European Union's (EU) General Product Safety Regulation (GPSR) is a set
of rules that requires consumer products to be safe and our obligations to
ensure this.

If you have any concerns about our products, you can contact us on

ProductSafety@springernature.com

In case Publisher is established outside the EU, the EU authorized
representative is:

Springer Nature Customer Service Center GmbH
Europaplatz 3
69115 Heidelberg, Germany

www.ingramcontent.com/pod-product-compliance
Lightning Source LLC
Chambersburg PA
CBHW050329070726
47750CB00003B/48